D. G. Robinson · U. Ehlers · R. Herken
B. Herrmann · F. Mayer · F.-W. Schürmann

Methods of Preparation for Electron Microscopy

An Introduction for the Biomedical Sciences

With a Foreword by
K. Mühlethaler

With 62 Figures

Springer-Verlag
Berlin Heidelberg New York
London Paris Tokyo

Prof. David G. Robinson, Ph. D.
Priv. Doz. Dr. Ulrich Ehlers
Prof. Dr. Rainer Herken
Prof. Dr. Bernd Herrmann
Prof. Dr. Frank Mayer
Prof. Dr. Friedrich-Wilhelm Schürmann

Fachbereiche Biologie und Medizin
der Universität Göttingen
D-3400 Göttingen

ISBN 978-3-540-17592-6 ISBN 978-3-642-48848-1 (eBook)
DOI 10.1007/978-3-642-48848-1

Library of Congress Cataloging-in-Publication Data. Präparationsmethodik in der Elektronenmikroskopie. English. Methods of preparation for electron microscopy. Translation of: Präparationsmethodik in der Elektronenmikroskopie. Includes index. 1. Electron microscopy – Technique. I. Robinson, David G. II. Title. QH212.E4P72 1987 502'.8'25 87-378.

2131/3130-543210

Foreword

In 1939, when the electron optics laboratory of Siemens & Halske Inc. began to manufacture the first electron microscopes, the biological and medical professions had an unexpected instrument at their disposal which exceeded the resolution of the light microscope by more than a hundredfold. The immediate and broad application of this new tool was complicated by the overwhelming problems inherent in specimen preparation for the investigation of cellular structures. The microtechniques applied in light microscopy were no longer applicable, since even the thinnest paraffin layers could not be penetrated by electrons. Many competent biological and medical research workers expressed their anxiety that objects in high vacuum would be modified due to complete dehydration and the absorbed electron energy would eventually cause degradation to rudimentary carbon backbones. It also seemed questionable as to whether it would be possible to prepare thin sections of approximately 0.5 μm from heterogeneous biological specimens. Thus one was suddenly in possession of a completely unique instrument which, when compared with the light microscope, allowed a 10–100-fold higher resolution, yet a suitable preparation methodology was lacking.

This sceptical attitude towards the application of electron microscopy in biology and medicine was supported simultaneously by the general opinion of colloid chemists, who postulated that in the submicroscopic region of living structures no stable building blocks existed which could be revealed with this apparatus. In those days one viewed the cytoplasm, the nuclear matrix and the mitochondrial and plastid interior as an amorphous, homogeneous gel-like scaffold, lacking defined structural delineation. Why then an electron microscope?

The lack of thin sectioning methods, which originally enabled a systematic investigation of the detailed structure of cells and tissue brought about a competitive search. As in the times of Antonie von Leuwenhoek, one worked with all possible and impossible objects, in the hope of observing something which would be both unique and epoch-making. Apart from the visualisation of viruses and bacterial flagella, this period gave rise to only a few real findings. Through the desire to publish immediately freshly taken pictures, many denatured structures and artifacts were interpreted in terms of living material, which obviously was not helpful to the application of this new method. As long as the use of electron microscopy was limited to the investigation of viruses, bacteria, erythrocyte ghosts, diatoms etc., no new information could be obtained. The preparation of these objects was without problems, since they could be suspended in distilled water and dried onto a collodion film. With the exception of diatom skeletons, most of the pictures were blurred and lacked contrast, thus hindering the visualization of more than just the outlines.

A metal shadowing method for the estimation of the depth of electron microscopical objects had already been developed in 1942 by H.O. Müller in Berlin, but surprisingly, this was not adapted by biologists for contrast improvement. Williams and Wyckoff, who independently developed metal shadowing for electron microscopy in 1944 in the United States, applied this technique to improve the representation of viral and bacterial preparations. The pictures obtained by these authors showed exceptional depth and represent a milestone in the development of electron microscopy, causing even the most sceptical of critics to become unsure. In 1946 Wyckoff was already using a freeze-drying method to prepare viruses and bacteria and thereby obtained an improved spatial structure preservation as compared to air drying.

Whilst very good results could be achieved with particle suspensions, thin sectioning was still required for larger objects such as cell walls, fibres, tissues etc. One applied all possible tricks in order to prepare sufficiently thin layers for microscopy, but the information which one thereby obtained was modest. We used, for example, ultrasonication to dissect plant fibres and, from the smallest fractured pieces, concluded a fibrillar composition which confirmed the earlier observations of Frey-Wyssling with the polarisation microscope. Instead of thin sectioning, replica methods were also used. The replicating materials used were gelatine or collodion, both of which were shadowed with SiO after removal from the object. It was, however, difficult, due to the expansion of the replica material, to remove the shadowed layer from the relief side without tearing it. The method of carbon coating introduced by Helwig and König in 1950 used sputtering and provided very good-quality replicas. Using this method it was possible to obtain excellent pictures of the sculptured exine surfaces of pollen grains.

The relief methods became of secondary importance, however, due to the development between 1948 and 1950 of reproduceable thin sectioning methods. The Americans Pease and Baker were successful in preparing the first such sections in 1948. They adapted a Spencer histological microtome 820 by reducing the knife advance to one tenth of its original value, and embedded the specimen in a mixture of collodion and Carnauba wax. The success obtained with this method stimulated other groups into further experiments with different microtomes and embedding materials. Bretschneider in Holland, for example, used paraffin with a melting point of 65° C as an embedding matrix and prepared his sections on an old Cambridge rocking microtome in a cold room at 10° C. His specimens were, however, poorly preserved structurally because he employed fixatives commonly used for light microscopy such as Bouin, Carnoy, bichromate-Formol or sublimate. A significant advance was achieved by modifying the microtome from a mechanical to a thermal advancing system. This was introduced by Newman, Borysko and Swerdlow in 1949. To do this, the specimen holder of a Spencer microtome was attached to a brass block containing a built-in reduction valve which could be cooled with CO_2. Moreover, a mixture of methyl and butyl methacrylate instead of paraffin or collodion was used as embedding material. A year later, Latta and Hartmann replaced microtome knives made of metal by the sharp cutting edge of splintered glass. This represented further significant progress in thin sectioning techniques. The only disadvantage of a continuous, thermally controlled advance system was that the specimen touched the knife edge again on the back stroke, which led either to deformation of the sample, or to damage to the knife. The simple up and down movement was therefore changed so that the knife arm could be moved in a circular manner, or the knife was retracted after each section to enable the specimen to pass by. Later, the microtome was driven by a motor in order to retain identical time intervals between the preparation of each section. The most commonly used apparatus at that time was the so-called Porter-Blum microtome which was first manufactured in 1953 by Sorvall.

After the establishment of thin sectioning methods, it became obvious that sufficient structure retention was not possible with the conventional fixatives, a fact already apparent in Bretschneider's investigation. The use of osmium tetroxide or Formol gave acceptable results, but produced variable structure retention, depending on the characteristics of the object. Palade discovered in 1952 that for an optimal fixation, the pH value always had to correspond to that of the object, and that even the osmolarity of the buffer was significant. At that time veronal buffer was almost exclusively used as a buffer system. Because of insufficient contrast in normal fixed and methacrylate embedded material it became standard practice to dissolve away the embedding matrix and to shadow the section prior to microscopy. During this procedure, however, the sections expanded considerably; this led to artifacts. This was changed in 1959 when Glauert introduced Araldite as a replacement for methacrylate (which vapourised under the electron beam). At roughly the same time it became general practice to increase the contrast of sections by staining them with lead salts.

Now that the technicalities of thin sectioning were solved, interest was directed towards the nature of the chemical composition of the structures in sections. Initially, one tried to find more information with the help of heavy metal staining or through enzymatic digestion, but despite tremendous efforts it was not possible to obtain convincing results. It became clear that for further progress in this direction one needed unfixed material to experiment with. It therefore seemed appropriate to develop a method for freeze sectioning material for histochemical investigations. We decided in 1958 to build such an apparatus, which required diverse constructional modifications of the conventional thin sectioning microtome. To avoid the accumulation of hoar frost on the frozen tissues during sectioning, the microtome had to be built into a vacuum chamber, which brought with it great technical problems. In collaboration with H. Moor and H. Waldner, a prototype was developed and tested in our laboratory. To our great disappointment we were not successful in preparing thin sections from frozen tissue. On the other hand, the shadowing of metal onto freshly cleaved surfaces enabled the production of good-quality replicas, as had already been demonstrated in 1957 by Steere with viral crystals. We obtained particularly interesting results on the membrane system of cells with the use of this freeze-etching technique. We showed that the structure of the "unit membrane", as postulated by Robertson in 1958, varied between different cell organelles. In particular, we noted that all of the biological membranes which we examined showed a particle population which should not exist according to Robertson's "unit membrane" concept. However, our observations were claimed to be artifacts by the experts of the day and it required many years until our work became accepted. In retrospect it is not unfair to say that the present knowledge of the structure and composition of biological membranes found its origin in our work and methods.

After all attempts had failed to prepare thin sections from unfixed frozen tissue, other approaches had to be taken to characterize the diverse structural elements biochemically. As shown by Coons, Creech and Jones in 1941, covalently bound fluorochrome tags could be used with antibodies to identify antigenic determinants in tissue sections in the light microscope.

It therefore seemed logical to use antibodies in the electron microscope, whereby the fluorescent tags were replaced with a heavy metal complex in order to visualise the marker molecules. As shown by Singer in 1959, ferritin-antibody conjugates proved suitable, due to the electron-dense iron core of ferritin. Even after shadowing, these particles could be clearly recognized. Over the years, markers other than ferritin such as peroxidase, hemocyanin, lectins or viruses found a use but they have all now been replaced by gold colloids. Gold particles can be prepared in a variety of sizes, which allows double labelling to be carried out on the same specimen. However, with chemically fixed preparations problems arise due to a modification or denaturation of the binding site. Thus in the last few years repeated attempts have been made to prepare frozen sections. The success achieved of late with cryomicrotomy has enabled significant advances to be made in immuno electron microscopy.

Unfortunately, even with the most sophisticated methods, it is not possible to approach the limit of resolution of the electron microscope, which is approximately 2 Å. Numerous reasons, such as denaturation, electron beam damage and optical factors, are to blame, and have not yet been completely eliminated. According to Siegel, for example, the crystalline structure in a paraffin crystal is changed despite the maintenance of the preparation at 4 K, even at a low dose of radiation of approximately 8 electrons/$Å^2$ at 100 kV. Similarly, Glaeser has shown that the crystalline structure of adenosine at room temperature is extensively destroyed after a dose of 6 electrons/$Å^2$ at 80 kV. Radiation damage also occurs at a dose far below the levels required for photographic recording. From this point of view the doubts expressed by cytologists at the time of the introduction of the electron microscope were not totally without grounds! These artifacts are not, however, as bad as once feared, since by conventional electron microscopy with a resolution between 30 and 30 Å the electron beam damage hardly influences the experimental results. It is, however, of significance that in the last few years a development in high resolution studies is in progress which is producing vital advances. On the one hand, one is trying to obtain better structural retention in the cryo-electron microscope with vitrified biological specimens and, on the other hand, one is now using computers in Fourier filtration and correlation analyses to eliminate artifacts. In particular, new image analysis methods have been developed such as morphometric stereological analysis and image reconstruction, which provide new information in regard to the structure and chemical composition of samples.

These new methods are certainly of great importance for high resolution electron microscopy, but cannot hide the fact that we are slowly approaching the resolution limit of the modern electron microscope which, as happened with the light microscope, hinders further developments. In every new area of research, there is a period of radical development (which in electron microscopy took place between 1950 and 1960) followed by a period of slower progress which is, however, simultaneously compensated by a broadening of the areas of application. In this respect, the development of the electron microscope progressed similarly to that of the light microscope, where its application in widely different areas has also followed specialisations in the apparatus (e.g. polarisation, phase contrast, interference contrast etc.).

The principles of the scanning electron microscope were worked out by Ardenne in 1933. However, the scanning microscope built by him in 1937 was originally constructed for transmission purposes since the chromatic aberration remained very small, even with thick specimens. An apparatus planned for depicting surfaces could not be perfected by Ardenne, due to the bombing of his laboratory in the war. Strangely enough it was not until 1965 that Oakley, Nixon and Pease in Cambridge built a surface scanning microscope. In the meantime the scanning electron microscope had entered various applicational areas with great success. By replacing the conventional cathode with a fixed emission source, Crewe managed in 1970 to elaborate on the original concept of the scanning microscope and produced a machine for high resolution purposes: the scanning transmission electron microscope (STEM).

The criteria for the preservation of structure in the scanning electron microscope correspond to those of the transmission electron microscope and it is therefore understandable that similar steps are involved in sample preparation. Amongst those methods especially useful for the structural preservation of the scanning electron microscope sample, the most successful has been the so-called critical point method, which Anderson developed in 1950 for the preparation of red blood corpuscles. In order to protect the sample from high thermal loads, the sample was in addition coated with a thin conductive metal layer. This sputtering method was introduced by Helwig and König in 1950 and became generally accepted. The films produced by this method adhere well to the surface and, even with the most sculptured reliefs, provide an evenly thick layer.

Those who, like myself, have witnessed since 1943 the whole of the development of electron microscopy, may have the satisfaction that all the years of toil and competition in the search for new methods of preparation and visualization have indeed paid off: our modern concepts of cellular structure exist mainly thanks to electron microscopy. The authors of this book on preparation methods for transmission and scanning electron microscopy are to be thanked for their excellent work. They have selected out of the 50 years of electron microscopy the most proven methods and presented them both clearly and lucidly. Despite the fact that the conventional methods in electron microscopy have remained more or less the same since the beginning, it requires considerable experience to select the most reliable methods from the numerous procedures available. For the beginner it is, therefore, a great help to use the advice summarised here in this book and spare the time-consuming burden of reading all the specialised literature. Even for the most experienced worker this represents a reliable reference book, enabling one to keep up to date with the most recent progress in preparation techniques. Despite a variety of books available on this theme, a complete collection of all the preparation methods has been lacking until now. It is therefore a stroke of good fortune that several authors, who each possess considerable experience in their specialised areas of research, have come together to produce this book. I am certain that it will become a standard book in laboratories and I wish this work a universal distribution and the authors the well-deserved acclaim for their tremendous work.

K. Mühlethaler, Zürich

Preface

The analysis of biological (and medically related) objects was only a side line in the pioneering days of electron microscopy. In the intervening years this aspect has developed tremendously, making it difficult to keep track of all of the various techniques. These advances are not just restricted to individual techniques but encompass nearly all steps in almost every technique. They stretch from developments in the preliminary preparation of objects and go through various fixation, embedding, staining and sectioning techniques right up to problems of image analysis. Nearly all of these developments have been associated with the construction of new machines or pieces of apparatus. These advances have, however, not been free from difficulties. As we reach the limits of resolution, both of the methods themselves and of the electron microscope, the artifact problem is no less important than it used to be. Nevertheless, things are still in a state of flux, particularly in those areas concerned with high resolution microscopy and element analysis.

There are relatively few laboratories engaged in the conception and development of new methods for electron microscopy. In contrast, there are numerous groups who have to apply daily standard methods on new objects. This occurs both for research as well as for diagnostical purposes. New methods must be tried out and new members of a group must be trained in both new and standard methods. This book is designed as a help to those involved in such a task. It describes successful methods which are regularly in use in the laboratory of one or more of the authors. Each section is prefaced by a short theoretical introduction and is concluded with citations of original literature and of books or review articles. Wherever possible, we have tried to indicate stages where artifacts can be expected and have tried to include an electron micrograph to demonstrate this. Some techniques which are still in the embryonic phase are mentioned, but not described in detail.

In the preparation of this book we have drawn heavily on the experience of our coworkers (both past and present). Although they were not directly involved in this undertaking, we would like to express our warmest thanks to them. Thanks are also due to numerous colleagues who encouraged us, who provided micrographs and who criticized our initial attempts. A special word of gratitude is due to Bernd Raufeisen, who finished excellent line drawings for this book, many of which were accomplished on the basis of simple sketches, and also to Heike Freundt and Bettina von Linde-Suden who helped greatly on the technical side during the preparation of the manuscript. Last, but not least, we would like to express our thanks to Dr. D. Czeschlik of Springer-Verlag for his patience, his readiness to help and for encouraging us to prepare this English translation of a text originally written in German.

Göttingen, March 1987 The Authors

Contents

1 An Introduction to Electron Microscopy (EM)

1.1 Imaging Methods in Electron Microscopy

1.1.1 Conventional Transmission Electron Microscopy (TEM)

A beam of electrons can interact with an object in a conventional transmission electron microscope in one of two ways. Usually elastic scattering takes place, whereby the electrons change their path in the specimen without a loss of energy. Inelastic scattering can also occur, resulting in a loss of energy due to an interaction of the impinging electrons with the orbital electrons surrounding the nucleus of each atom in the object. Those electrons which are not or hardly scattered contribute positively to the image. Those which are considerably deflected are prevented from doing so by apertures in the optical path. As a result differences in light intensity (contrast) are created in the final image, which relate to areas in the object with different scattering potentials. This fact can be deduced from the following formula of Rutherford (adaptation of Coulomb's law), which describes the deflection potential of an atom:

$$k = \frac{-e \cdot eZ}{r^2},$$

in which k = deflection potential, e = electron charge, Z = positive charge of the nucleus (atomic number), r = distance from electron to nucleus.

1.1.1.1 Bright Field Electron Microscopy

The electron beam is hardly deflected by those elements (e.g. C, O, H, N, S, P) which are present in biological material. Thus biological objects have little or no inherent contrast when viewed in the electron microscope. Instead, contrast has to be obtained artificially by introducing elements with higher atomic number (e.g. Os, Mn, U, Pb, Pb, W etc.) into the object. This can be achieved during the fixation (e.g. with OsO_4, see Chap. 2.1.1.2.1 or $KMnO_4$, see Chap. 2.1.1.2.3) or through special staining (see Chap. 2.3.1.3 and 2.6.5) techniques.

An aggregation of atoms with low atomic number can sometimes be seen in the TEM without staining. The mass thickness or density of the atoms at the region in question is then obviously such that a scattering of the electrons in the beam does, in fact, occur. Clearly such an effect will take place more and more frequently the thicker the object becomes. As a result the maximal thickness of a biological specimen in a TEM, operated at normal (100 kV or less) accelerating voltages, lies around 500 nm. At greater thicknesses the electron beam will penetrate but cannot effectively leave the object. Hence in order to produce meaningful micrographs a biological object in a TEM should not exceed 100 nm in thickness. This fact constitutes one of the major problems in object preparation for biological TEM.

Contrast is modulated by two other factors. One is the size of the objective aperture: the smaller the opening becomes, the greater is the contrast achieved (at a loss, however, of beam intensity and resolution). The second factor is the accelerating voltage which is applied between the cathode and anode: the greater the voltage becomes, the greater is the kinetic energy of the electrons. This in turn causes a reduction in the angle of their deflection and therefore a reduction in contrast.

Literature

Agar AW, Alderson RH, Chescoe D (1974) Principles and practice of electron microscope operation. In: Glauert AM (ed) Practical methods in electron microscopy, vol II. Elsevier/North Holland, Amsterdam, pp 1-345

Lange RH, Blödorn J (1981) Das Elektronenmikroskop. TEM und REM. Thieme, Stuttgart, pp 1-327

Wischnitzer S (1970) Introduction to electron microscopy. Pergamon, New York, pp 1-292

1.1.1.2 Low Dose Transmission Electron Microscopy

Under routine conditions in normal TEM the specimen is exposed to an electron beam intensity equivalent to several hundred to several thousand electrons per Å^2. The total dosis is composed of the individual doses which correspond to searching, focussing and exposing the specimen. One can, however, avoid beam damage to the specimen by reducing this total dose. Some EMs have built-in devices for this to be done on a regular basis. The procedure is as follows: scan the specimen at the lowest possible magnification and beam intensity; the dose will be around 10^{-2}e per Å^2 s^{-1}; focus at the desired magnification, but not at the position which is to be exposed; move the specimen to the position required (during this step one should try to shield the specimen from the beam); finally expose the specimen with the first electrons which impinge on the region in question at the desired magnification. Done in this manner, negatives can be obtained from a specimen, or part of it, which has been exposed to only 6-10 electrons per Å^2 as the total dose. Naturally in order to obtain the maximum of information one should use a special film emulsion and an appropriate developer. As a check on the quality and correctness of information obtained one usually takes a second picture of the same region but under routine conditions (the focus remaining unchanged).

Determinations of the electron dose are not possible in EMs which are not properly equipped for this purpose or when the recommended procedure is not exactly followed. A rough estimate can be made provided that a highly sensitive film emulsion is used and one develops the film correctly. Under these conditions a total dose of between 5 and 50 electrons per Å^2 leads to a darkening of the negative at magnifications of between 20,000 and 40,000 ×. This procedure does not, of course, take into account the exposure of the object to the electron beam during the initial scanning of the specimen.

Electron beam damage can be demonstrated on crystalline objects with the help of light optical diffraction analysis, whereby the peripheral diffraction points/arcs disappear with increasing electron dosis. Thus a loss in resolution is registered which can be quantitatively expressed (Fig. 1.1 a–e).

Literature

Unwin PNT, Henderson R (1975) Molecular structure determination by electron microscopy of unstained crystalline specimens. J Mol Biol 94: 425–440

Wasserbäch W (1983) Kontrast- und Strahlenschädigung in biologischen Präparaten. Zeiss Inf Mem 2: 28–41

Wrighley NG, Brown E, Chillingworth RK (1983) Combining accurate defocus with low-dose imaging in high resolution electron microscopy of biological material. J Microsc 130: 225–232

Wurth M (1983) „Low-dose"-Mikrographien des Bakteriophagen T4 – aufgenommen mit dem Zeiss EM 109 Elektronenmikroskop in Mikro-Dosis-Technik. Zeiss Inf Mem 2: 25–27

1.1.1.3 Dark Field Electron Microscopy

Usually TEMs are equipped with a beam-tilting device which is necessary for dark field microscopy. With this the electron beam impinges upon the specimen at an angle of less than 90°. Thus only the strongly diffracted electrons can contribute to the final image, whereas the great majority of scattered and nonscattered electrons is trapped by the object aperture and is removed from the optical path.

Holes in a film appear black in dark field as compared to the object which is more or less bright. This means that only clean, thin support films can be used if the object is to retain this difference in contrast. A typical example for the application of dark field microscopy is the visualisation of BAC-spread nucleic acids (see Chap. 2.3.2.4), which are either observed in unstained or stained (with urany lacetate) form.

Dark field microscopy requires a high electron beam dosis which can lead to an incineration of the specimen. Moreover, contamination of the specimen, which acts like a thick support film, must be prevented. This is usually achieved with a specimen stage cooling device. Focussing requires, as a rule, more experience than in bright field microscopy, since aids to focussing such as phase contrast features of the background can no longer be used. Difficulties are met

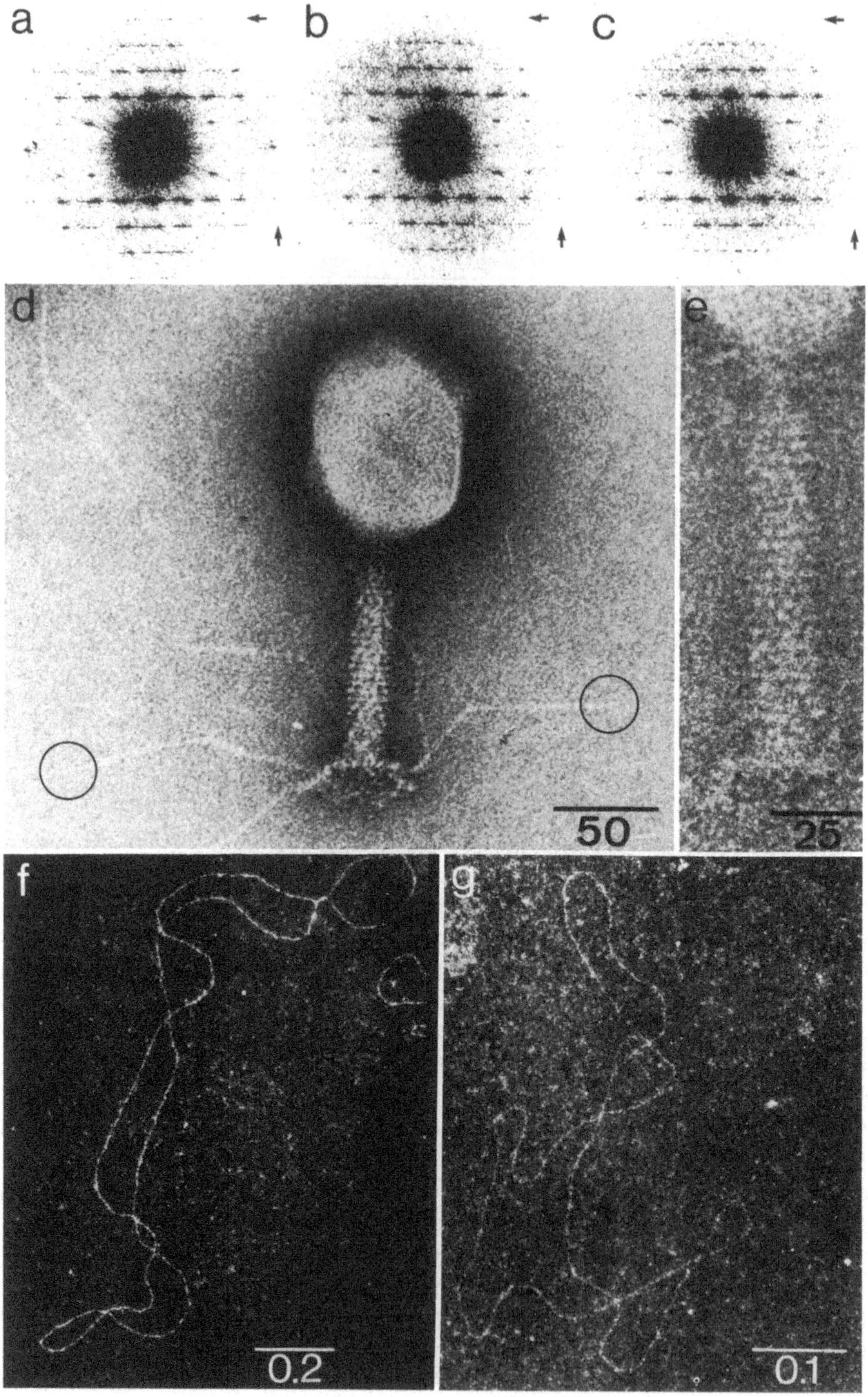

Fig. 1.1 a–g

with the exposure times since, with a bright object against a dark background, the normal exposure meter cannot always be used.

Although the image quality is poor, dark field conditions can also be achieved by displacing the object aperture. To the same end one can also replace the condensor aperture by a condensor ring aperture. The latter method requires the precise realignment of exactly corresponding sets of apertures and has not found much following in biological laboratories.

1.1.2 Conventional Scanning Electron Microscopy (SEM)

1.1.2.1 Imaging with Secondary and Back-Scattered Electrons

Although the general principles of elastic and inelastic scattering also apply to electron beam-specimen interactions in SEM, their relative contributions to the final image are different to that in TEM (see Fig. 1.2). Elastically scattered electrons are reflected back from the surface or surface layers of the specimen. Inelastic scattering occurs as a result of an interaction between the electron beam and atoms within the specimen. In this case so-called secondary electrons, with less energy than the impinging electron beam, are emitted from the specimen. Both secondary and backscattered electrons contribute to the image in SEM. Two further electromagnetic signals arising out of the interaction between electron beam and specimen which are of interest for elemental analysis (see Chap. 3.1.1) are X-rays and cathodoluminescence.

The emission of secondary electrons is restricted to a thin surface layer of the specimen. The thickness of this layer is determined by the energy of the secondary electrons and is independent of the energy of the impinging electron beam. For biological materials this layer reaches 10 nm but for heavy metals it seldom exceeds 1 nm. This feature forms the basis of resolution in SEM.

The number of secondary electrons which are emitted rises with increasing angles of incidence between the impinging beam and the specimen. However,

◁

Fig. 1.1. a–c Effect of beam damage on negatively stained actin layers. Light optical diffractograms have been made from micrographs prepared with the following electron doses: **a** 50 $e^-/Å^2$; **b** 150 $e^-/Å^2$; **c** 250 $e^-/Å^2$. High order reflections are indicated with an *arrow* in **a**, become weaker in **b** and are completely absent in **c**. This indicates that the crystalline structure of the object is partially destroyed
d, e Structural preservation through low dose EM as demonstrated on the T4 bacteriophage (**d** 8.5 $e^-/Å^2$; **e** 7 $e^-/Å^2$). Details *(in circles)* are usually not recognizable at normal (high) electron dose. **a–c** from Wasserbäch (1983) (Micrographs W. E. Foster and U. J. Aebi); **d** from Wurth (1983) (Negative staining with uranyl formate); **e** negative staining with uranyl acetate. Magnification bar in nm
f, g Dark field micrographs of PM 2-DNA molecules (double-strand bacteriophage DNA, BAC-spreading technique, thin carbon film on a holey support film, see also Fig. 2.20); **f** positive staining with 0.1 mM uranyl acetate in acetone; **g** unstained, i.e. inherent contrast with BAC. (Original micrograph U. Hahn). Magnification bar in µm

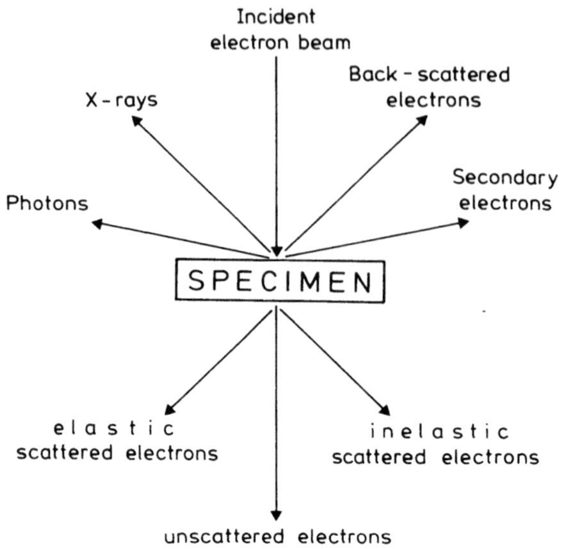

Fig. 1.2. The major electromagnetic signals which are emitted when an electron beam interacts with an object. Those signals sent back from the object are important for REM, those leaving the other end of the object are exploited in TEM. For details see text

the number of electrons which are finally registered by the detector is dependent upon a number of other factors: the voltage potential at the collector, the angle of tilt between specimen and detector, the morphology and charge accumulation at the surface of the specimen and of course the chemical nature of the specimen being investigated.

The positively charged collector causes the secondary electrons to assume a curved pathway. Thus electrons can reach the detector from regions in the specimen which do not lie perpendicular to the plane of the detector. Secondary electron images therefore possess great internal contrast; however, they have the disadvantage of being relatively "flat" (Fig. 1.3).

Backscattered electrons are not affected by the positive potential at the collector, although their emission and detection is to a large extent subject to the same factors as for secondary electrons. They too contribute to the final SEM image and are responsible for shadow contrast and therefore for the quality of plasticity in the image under normal operating conditions. Pure backscattered images can be obtained by reducing the collector potential. These have an extremely high contrast (Fig. 1.4), which can often be of use in structure determinations, but they lead to a much poorer resolution and there is a lot more "noise" in the signal at magnifications greater than 1000 ×. For biological objects the conventional SEM image (Fig. 1.5) is derived principally from secondary electrons. At lower magnifications, however, the relative proportion of backscattered electrons can be increased according to the need for more morphological details. One notes (see literature) that there are other methods of imaging in SEM but these are, as a rule, not important for biologists.

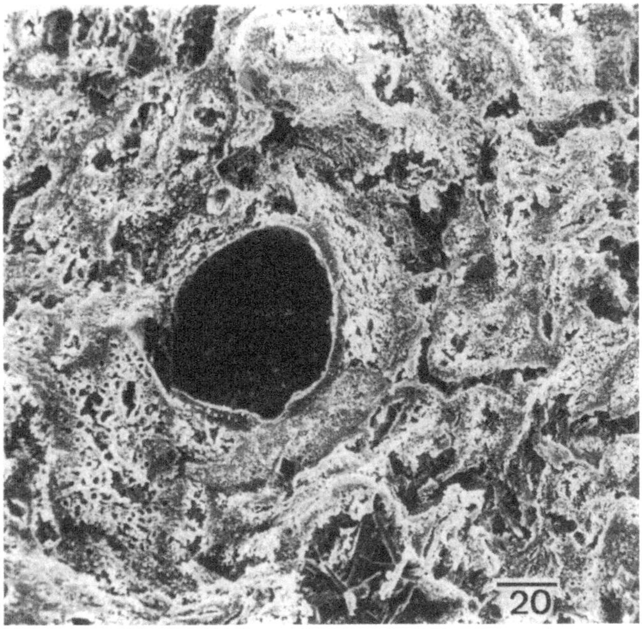

Fig. 1.3. REM-secondary electron micrograph. Compact bone of the femur from a Le Moustier Neanderthal individual *(Homo sapiens neanderthalensis)* after decomposition and remineralisation. Magnification bar in μm

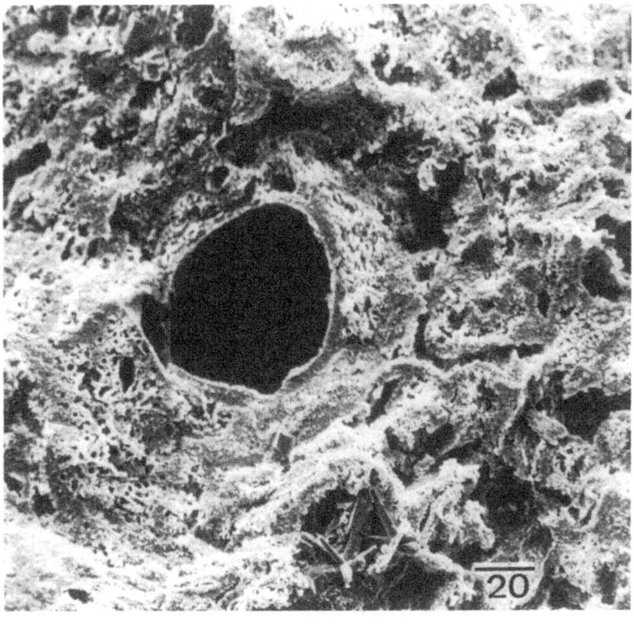

Fig. 1.4. REM micrograph from backscattered electrons. Other details as in Fig. 1.3

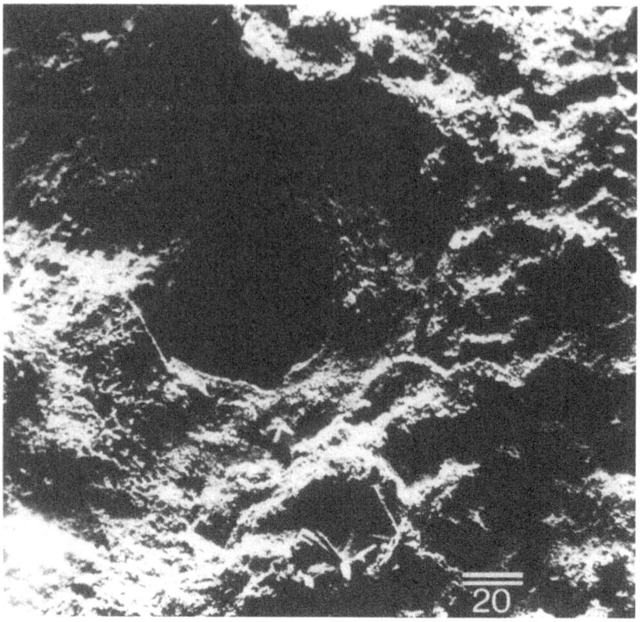

Fig. 1.5. REM micrograph from secondary and backscattered electrons. Other details as in Fig. 1.3

Literature

Goldskin JI, Newbury E, Echlin P, Joy C, Fiori C, Lifshin E (1981) Scanning electron microscopy and X-ray microanalysis. Plenum, New York London, pp 53-122, 172-182

Lange RH, Blödorn J (1981) Das Elektronenmikroskop. TEM und REM. Thieme, Stuttgart, pp 54-63

Reimer L, Pfefferkorn G (1977) Raster-Elektronenmikroskopie. Springer, Berlin Heidelberg New York, pp 34-55

1.1.2.2 Scanning Electron Microscopy at Low Accelerating Voltages

There are a number uf undesirable side effects (charge accumulation (see for example Fig. 1.6), warming up and the incineration of organic material) when the electron beam interacts with the specimen, and these are usually dependent upon the accelerating voltage of the electron beam. In some cases gases may be released from the specimen which lead to quadratic halos surrounding the scanned area. Such artefacts can be avoided by reducing the accelerating voltage. This, of course, causes a reduction in resolution but has the advantage that nonconducting specimens can also be investigated at low magnifications.

Biological objects are poor electrical conductors. They are usually coated with a thin layer of conducting material before being inserted into the SEM. This is, however, not always possible, particularly when the specimens are irre-

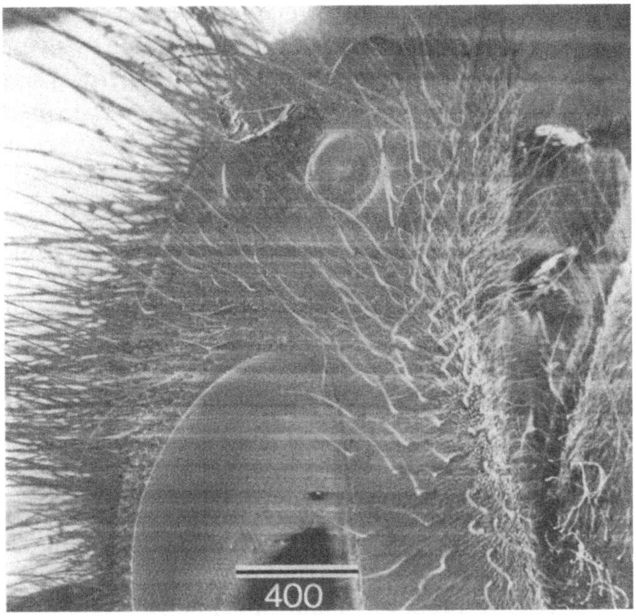

Fig. 1.6. REM micrograph of a wasp's head taken at an accelerating voltage of 2 kV (untreated, museum specimen). Despite a relatively small area of exposure and the uptake of moisture before exposure to the electron beam there is still appreciable surface charge accumulation. Magnification bar in μm

Fig. 1.7. REM micrograph of a daisy (*Bellis perennis*, fresh specimen, untreated) taken at an accelerating voltage of 1.5 kV. Magnification bar in μm

placeable, e.g. museum objects. In these cases low dose (=low voltage) conditions are obligatory. It may also be desirable to investigate objects in their native state such as fresh plant or animal tissues. However, specimens of this type have to be pretreated to avoid volume changes. The simplest way of doing this is by placing the specimens in an exsiccator at low vacuum for 1–2 days followed by a slow recovery to atmospheric pressure (see Chap. 3.5 for more sophisticated procedures). The resulting specimens are not dry, and the moisture in them is advantageous since it reduces the tendency to surface charge accumulation. Indeed it is recommended that dry, e.g. museum, specimens absorb moisture from the air (or if need be by placing them is a closed container with water) before investigating them under low voltage conditions. Mounting fresh specimens poses no great problem: two-sided sticky tape, silver or copper conducting paint or glue (see also Chap. 3.6) can be used. For some objects, e.g. pinned insects, it may be necessary to construct some kind of aluminium holder.

In order to reduce the length of exposure to vacuum conditions the specimens should be oriented with the help of a binocular microscope before inserting them into the SEM. Observations are then made at magnifications around 500 × with accelerating voltages between 1.5 and 2 kV. Under these conditions even freshly prepared plant specimens (Fig. 1.7) usually remain stable long enough for scanning and photography to be completed, and by electrically modulating the video signal, one can also compensate for strong changes in contrast which may set in with time.

Actually, the advantageous features of low voltage SEM are not without grounds for concern. Clearly the introduction of unfixed, soft, water-containing specimens into a vacuum system is a strain for the pumps and leads to contamination of the column with water vapour and other volatile components. Secondly there are now binocular microscopes available which are so good that almost everything that can be seen with SEM under low voltage/magnification conditions can also be seen with a light microscope. Finally, there are now methods available for the removal of metallic conducting coats (e.g. Au, see Chap. 3.7) so that specimens can easily be brought back, undamaged, to their original form. Thus one should weigh up the advantages of the SEM operating at low voltage conditions very carefully.

1.2 Preparation Procedures in TEM

1.2.1 Overview

Except for certain cryo methods, specimens for TEM are of necessity water-free in order to withstand the high vacuum conditions to which they become exposed. In general there are two ways of preparing biological objects for TEM:

a) *Indirectly.* In this case a replica of the object, but not the object itself, is brought into the electron beam. One differentiates between replicas of more

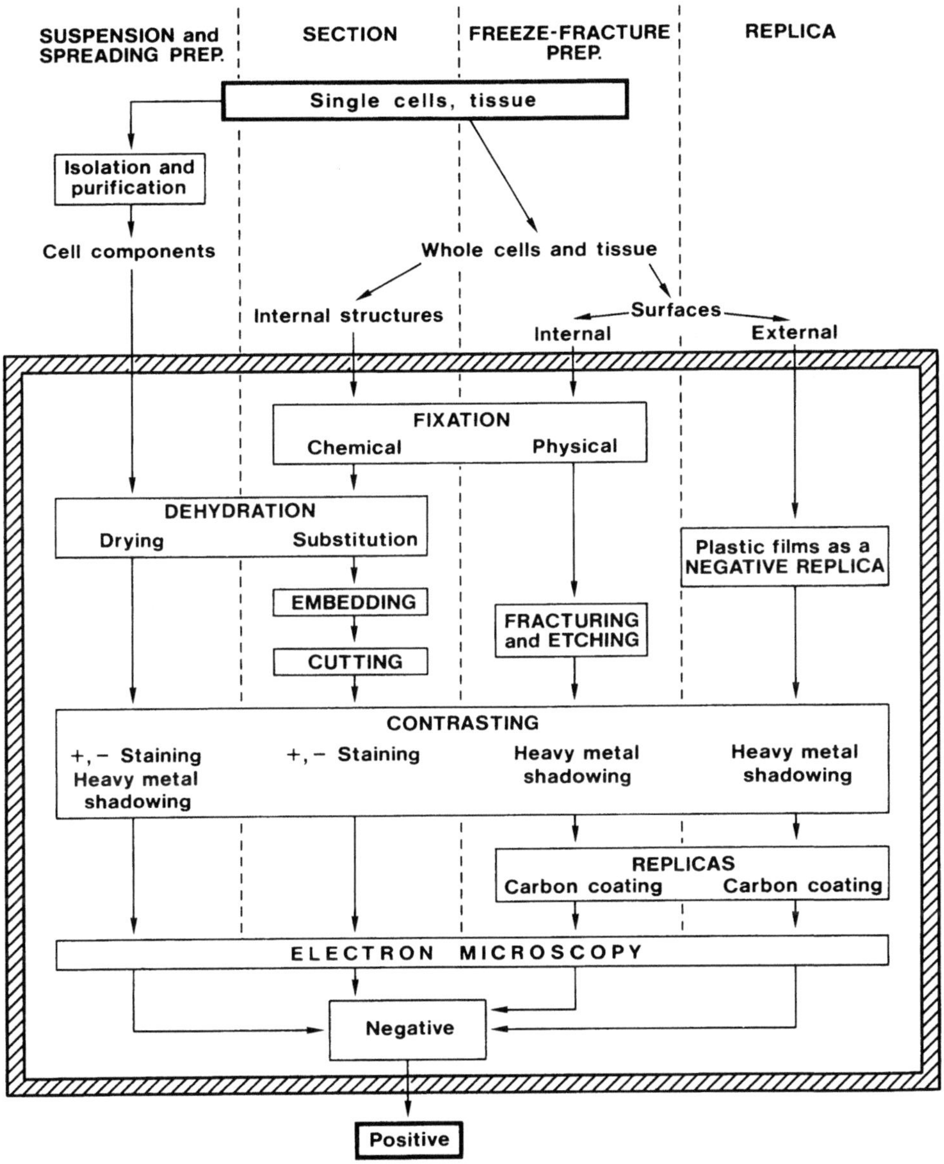

Fig. 1.8. Methods of preparation in TEM: an overview

or less dry *external* surfaces from those of *internal surfaces*. For the former, fixation and dehydration steps are not necessary in contrast to the latter where the object must be subjected to a physical fixation through quick freezing (freeze-etch method, see Chap. 2.6).

b) *Directly*. Here the object is brought into the electron beam and therefore must be dehydrated beforehand. Depending on their size, one differentiates between suspension preparations and sections. In the former the dehydration is achieved without a substitution, e.g. by drying in air. With the latter, whereby cells or tissues are cut into *sections* thin enough for the passage of an electron beam, dehydration is coupled to the impregnation with an embedding substance.

The relationships between the individual preparatory procedures and the various staining possibilities are presented diagrammatically in Fig. 1.8.

1.2.2 Structural Preservation During Fixation, Dehydration and Embedding of Biological Objects

A chemical fixation which acts to stabilize the various cellular structures must be undertaken on objects which are to be sectioned. Fixations are necessary:

- in order to prevent cell lysis occurring after organismal death,
- in order to prevent the loss of a variety of substances during dehydration,
- in order to avoid possible changes in structure due to staining with heavy metals.

Ideally, in a good fixation all structures should be optimally preserved. In particular one hopes that, after a fixation, the chemical reactivity (for histochemical purposes) has been changed as little as possible. Theoretically all structures should be fixed to the same extent, but in practice one finds that fixatives tend to act selectively. A way out of this dilemma is to use a combination of various fixatives and to compare the results from different fixations, including freeze fixation. There are in general two types of fixatives:

a) *Coagulatory.* These denature proteins and cause them to precipitate as large granular or fibrillar aggregates which are in suspension. Since there are no bonds formed between the various molecules this kind of fixation is also termed nonadditive. Examples of this kind of fixatives are: ethanol, methanol, HCl, chromic acid.
b) *Noncoagulatory.* Fixations of this type change the sol form of the cytoplasm into an electron transparent gel. They do so by creating a network of bonds between the various molecular components and are therefore „additive“ in function. Examples of this kind of fixative are OsO_4, $KMnO_4$, glutaraldehyde. Only additive fixatives are suitable for electron microscopy.

The quality of a fixation is determined by a number of factors, the most important of which are the concentration and the rate of penetration of the fixative, the temperature and the duration of the fixation, and lastly the size of the object. These factors are also related to one another; for example, the concentration influences the duration of the fixation; the temperature influences the rate of diffusion of the fixative and therefore the duration of the fixation; the penetration rate determines the size of the object to be fixed. As a rule, and irrespective of the temperature and duration of the fixation, one does not fix objects which are larger than 1 mm^3 since the homogeneity of the fixation cannot be guaranteed.

Since the majority of embedding media are insoluble in water and because EM specimens dare not contain water, the fixation is followed by a dehydration. Usually this is achieved by replacing the cell water by an organic solvent. The solvent chosen should possess the following properties: ease of penetration, poor capacity for extracting cellular constituents, poor reactivity with fixatives, good miscibility with embedding media. Acetone fulfills all of these requirements, ethanol all but the last. In this case a transition solvent, e.g. propylene oxide (see Chap. 2.1.2) must be used.

A number of requirements have also to be met by the embedding medium:

- it should penetrate quickly and quantitatively into the object (the viscosity of the embedding medium is proportional to the size of the monomers and low viscosity is advantageous);
- it should be inert as far as the cell and its constituents are concerned;
- it must polymerize (harden) within a reasonable length of time and at a temperature of less than 100° C without changing in volume; the polymerization should occur homogeneously to produce an "even" block and should be reproducible;
- it must allow thin sectioning;
- it must not present a barrier for staining solutions, nor should it become stained itself;
- it must remain stable in the electron beam;
- it should be essentially electron-translucent;
- it must remain stable at high vacuum, i.e. it must not contain volatile components;
- it must not be hygroscopic;
- it should, if possible, not be toxic.

These requirements are, for the most part, fulfilled by the epoxy resins (see Chap. 2.1.3).

1.3 Imaging Problems

1.3.1 On the Interpretation of TEM Images

It is clear from the foregoing that the images obtained in conventional TEM are sui generis artefacts. This is particularly so for replicas where the object itself is not inserted into the electron beam. As a rule what we "see" is essentially a pattern of the distribution of heavy metal atoms which have been added secondarily to the object. Unfortunately, the question as to whether an EM micrograph really does represent the true life situation cannot be effectively answered: one is forced to collect information using a variety of methods. As a consequence of this a particular interpretation can be made more likely, but never certain.

Numerous possibilities exist on the way from object to negative where artefacts can arise. Almost each preparatory step has potential pitfalls:

a) During a chemical fixation (see Chap. 2.1.1)
- chemical reactions can occur which may lead to changes in conformation and reactivity;
- due to the selective nature of a fixative, substances can be washed out during a fixation;
- since fixations are not instantaneous there is always the possibility of cellular changes through cytolysis after cell death;
- since fixations are not always isotonic with the cell fluid and membrane permeability can be changed during a fixation there is a distinct possibility that a swelling or shrinkage may occur.

b) During a physical fixation (see Chap. 2.6.2)
- deformation may occur through ice crystal formation;
- phase separations may occur.

c) During dehydration (see Chap. 2.1.2)
- drying a suspension of organelles or particles in air can lend to their deformation (considerable pressures can develop at the surface of a drying object).

d) During embedding (see Chap. 2.1.3)
- deformation may occur as a result of volume changes during the polymerization.

e) During microtomy (see Chap. 2.2.3)
- considerable shear forces are developed which can lend to the displacement of cells and organelles;
- vibrations may occur leading to so-called chatter;
- the object can become warmed up due to the frictional events taking place at the specimen-knife interface.

f) During fracturing and etching (freeze etching see Chap. 2.6.3 and 2.6.4)
- mechanical deformations can occur, e.g. knife marks, ruptures etc.

g) During positive and negative staining (see Chap. 2.2.5.2 and 2.3.1.2)
- structural details can become masked by the presence of too many heavy metal atoms or by aggregates of the same. Unspecific precipitation of staining molecules can also occur.

h) During shadowing (see Chap. 2.3.1.3 and 2.6.5)
- objects become unavoidably larger due to the thickness of the coating material;
- the morphology of objects can be changed due to the deposition of crystalline aggregates.

i) During exposure to the electron beam
- charge accumulation (ionisation) and thermal changes leading to the contraction of both support film and section can occur;
- two types of specimen contamination can occur:

1. hydrocarbons (coming from the oil in the vacuum system) can condense on the specimen and become oxidized to carbon which forms a film and leads to a "graininess" in the specimen;
2. water vapour can condense on the specimen and reacts with cell constituents. As a result CO and CO_2 are liberated and this leads to a "bleaching" of the specimen.

Avoiding or reducing the artefacts is one of the principal tasks with which an electron microscopist is faced. A control or measurement of the success of one's efforts in this regard is only possible through a comparison of the results obtained with different methods on one and the same organism. This also includes light microscope data and information obtained from other physical or from biochemical methods. Continual progress in methodology in the biosciences means that our concept of the cell and our knowledge of its constituents are subject to constant revision.

Finally, it is important to stress the fact that statements about dynamical aspects of the cell are very difficult to make from electron micrographs, which in essence are only momentary "stills". Only indirectly, e.g. through pulse-chase experiments (see Chap. 2.5), can one obtain information on cellular events such as transport and motility as well as cellular development.

1.3.2 On the Interpretation of SEM Images

As discussed in Section 1.1.2.1, the type of image obtained in the SEM is dependent upon the signal (secondary/backscattered electrons) received and the way in which it is analyzed. The backscattered image probably reflects the true situation more than the secondary electron image, which contains more than just morphological information. Since secondary electrons are emitted from a layer at the object surface, they also possess information about the chemical nature of the object. Thus heavy elements appear brighter in the SEM than do those with a lower atomic number. In addition, elevations are registered more brightly than flat regions. The variable geometry of the system, i.e. the relative positioning of the object with respect to the electron beam and the detector together with the ability to electrically moderate the video signal enable one to obtain a multitude of different images.

In principle, the same problems and mistakes which were described in Section 1.3.1 for TEM can also occur in SEM when the preparatory step is the same. This also applies to the behaviour of the specimen in the electron beam, although heat development is a minor problem in SEM. There are, however, a number of special features of the technique which are particularly relevant to image interpretation in SEM. The most important of these lies in bestowing electrical conductivity to the object, which, as a rule, determines the quality of the image obtained.

An increase in the accelerating voltage of the electron beam is necessary for an increase in resolution, but this brings with it a greater tendency for charge accumulation at the surface of the specimen. High accelerating voltages enable the electrons to penetrate a thicker layer at the surface of the object and thus increase the relative number of secondary electrons which are emitted. Providing the object is small or thin enough, this feature can lead to the electron beam traversing the entire specimen. This, of course, enables one to obtain even more structural information on the three-dimensional appearance of the object. The only other way to increase resolution, namely through a reduction in working distance, leads inevitably to a loss in depth of focus.

1.4 Support Films

1.4.1 Grids for TEM and Their Pretreatment

There are a multitude of different grids for TEM. They are usually made of copper, although gold and nickel are also in use. They have a diameter of 2 or 3 mm. Basically there are two types of grid: those with a single hole or slot which can vary in form and size and those having a meshwork of systematically arranged holes or squares. One chooses a grid according to the appropriate method of specimen preparation. Thus, for negative staining, heavy metal shadowing and nucleic acid preparations, small square mesh grids are recommended, for thin sections large square mesh grids and for ribbons of serial sections slot grids.

For freeze-etch replicas various grid types can be used depending on the size of the replica and whether or not the grid has a support film or not. Finally there are so-called index or finder grids whose mesh system is numbered or lettered to allow for easy orientation in the electron beam and hinged or double grids. Cu-grids are usually clean enough to be used directly upon purchase but should there be doubt about this they should be rinsed in acetone. Grids have a rough and a smooth surface; the support film attaches better to the former. In order to improve this attachment (especially in the case of carbon-coated grids), one dips the grid into either a solution of neoprene (0.5%, w/v in toluene) or a solution prepared from sticky tape (tape immersed in toluene). The excess solution is dried off with filter paper.

1.4.2 Formvar Films

Formvar is polyvinyl formaldehyde. This substance, from a solution, dries to a thin film on smooth surfaces such as glass.

Procedure for Grid Coating: A solution of 0.5% w/v formvar in water-free chloroform (or dioxane or dichloroethane) is prepared and poured into a dropping funnel. A clean microscope slide is then warmed over a heating plate and placed into the formvar solution. The solution is then allowed to leave the funnel under gravity and the film of formvar remains on the slide. Three parameters determine the thickness of the film: the concentration of formvar in the solution (the more concentrated it is, the thicker the film), the speed with which the solution is allowed to leave the funnel (the quicker it is, the thicker the film) and the length of time during which the slide is allowed to remain in the funnel after the solution has left (the shorter this is, the thicker the film). The slide is then removed from the funnel and dried on the heating plate. It can then be stored in a covered beaker.

In order to release the formvar film from the slide, one takes a clean scalpel or razor blade and scores a rectangular form (about 2×4 cm) on the slide. The slide, with the scored surface upwards, is then slowly inserted at an angle into a trough of distilled H_2O. The rising water meniscus pulls the formvar film away from the slide at the scored edges. The formvar film finally floats off onto the surface of the H_2O. Those films which are grey and homogeneous are usable.

Grids are then placed with their rough surface onto the film. One takes an appropriately cut piece of dust-free, rough paper or the backing of parafilm and lays it carefully onto the film with grids. Finally the film-grid-paper sandwich is removed from the water with tweezers and laid onto filter paper with the film upwards. Filmed grids should be left to dry and stabilize for at least 12 h in the dark. If necessary, they can be reinforced with carbon as described in Section 1.4.6.

1.4.3 Collodion Films

Collodion is nitrocellulose; it is soluble in isoamyl acetate. Films can be prepared as described for formvar (see Sect. 1.4.2) or by the drop method.

Drop Method: Allow a drop of 3.5% collodion (w/v, in water-free isoamylacetate) to spread on the surface of distilled water. After evaporation of the solvent a thin layer of collodion is left on the water surface. If the water surface is not clean (presence of dust particles etc.), this film can be discarded (by picking up with tweezers) and another made in its place immediately afterwards.

Grids are added to the film and then picked up as for formvar (see Sect. 1.4.2). Alternatively (and this also applies to formvar films) grids can be carefully laid on a piece of filter paper which in turn lies on a metal mesh at the bottom of the water trough. The water is then slowly released from the trough and at the same time the film is carefully guided into position onto the grids. The mesh plus paper plus filmed grids is then lifted out of the trough and left to dry overnight as above.

Literature

Spiess E, Mayer F (1976) Herstellung einer Kunststoffträgerfolie (Film C 1189) Wissenschaftlicher Lehrfilm der IWF Göttingen (Institut für den wissenschaftlichen Film, Nonnenstieg, D-3400 Göttingen) aus der Reihe Elektronenmikroskopische Präparationsmethoden (also available in English)

1.4.4 Hydrophilisation of Films

Support films tend to be hydrophobic. To make them hydrophilic one employs the following procedure: ethidium bromide (5 mg ml^{-1}) is dissolved in double-distilled H_2O and centrifuged at 160,000 *g* for 12 h to remove contaminating particles. The supernatant is carefully removed and diluted with double-distilled H_2O to 50 mg ml^{-1}. A drop of this solution is then placed on a piece of parafilm and a filmed grid laid down onto the drop (film surface downwards). Leave for 10 min in the dark at room temperature, then wash three times (5 s) in double-distilled H_2O. After being dried on filter paper the grids retain their hydrophilicity for several hours.

An alternate method to achieve hydrophilicity of support films is by glow discharge. This is produced by slowly introducing air into a specially equipped evaporating unit. The filmed grids are usually mounted onto a copper support before the discharge and should be used within hours of their hydrophilisation.

1.4.5 Support Films with Holes

If extremely thin support films are required, particularly those of carbon, the square size in the grid mesh may be too large for adequate stability to be maintained. Instead one uses a "holey" support film, i.e. a support film which has many small holes, which is then coated with carbon. The support film thus assumes the role of a "mini-mesh" and supports the carbon film (see Fig. 1.9).

Preparation of "Holey" Films: A procedure recently developed by Lünsdorf and Spiess which produces reliable and consistent results is as follows:

a) Wash the collodion with ethanol and dry it carefully under an electric light bulb (60 W, distance 20 cm, 5 min). Then dissolve 0.1 g in 75 ml of ethyl acetate by stirring overnight in a brown glass-stoppered flask at room temperature.
b) Add 125 µl of aqueous 10% (w/v) BRIJ 58 and 0.92 g 87% (v/v) glycerol to the collodion solution.
c) Shake vigorously by hand for 2–3 min.
d) Immediately pour the turbid solution into a glass beaker. Dip a freshly cleaned glass slide into the solution for 10 s and after slow withdrawal allow to dry for 10 min by standing upright in a beaker. Factory pre-cleaned slides can be used without further pretreatment.

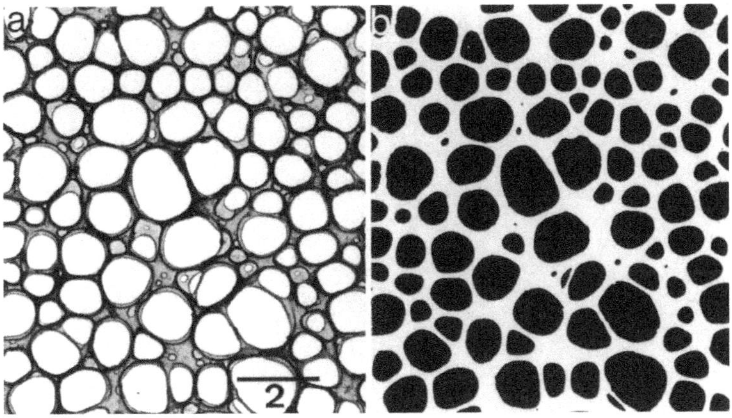

Fig. 1.9 a, b. Micrographs taken of a holey film. **a** bright field conditions; **b** dark field conditions. Magnification bar in μm

e) The perforated supporting films are separated from the microscope slides by floating the foils on double-distilled water in a glass Petri dish (14–20 cm diameter; 4–7 cm height). The slide should be dipped slowly into the water at an angle of 30°–40°.
f) Grids (200–300 mesh) are placed on the film with the rough side facing the film. The grids (maximal number: 35–40) should be placed in the centre of the film, taking care not to come too close to the borders. The film-covered grids are removed from the water surface with a piece of filter paper. They are then placed on filter paper in a Petri dish and air-dried at room temperature.
g) The dry film-covered grids are baked in an oven at 180° C for 10 min. This causes the areas covered by the collodion film to perforate into a net (holey film).
h) The film is subsequently covered by a thick layer of carbon (resistance evaporation in a vacuum of 6×10^{-4} torr) to achieve mechanical stability.
i) In order to obtain good adsorption of specimens the grids are glow discharged in an intensive plasma glow for 10–25 s.

1.4.6 Carbon Films

Carbon films are made by evaporating carbon [actually it is more correct to speak of a carbon sublimation, but in order to avoid confusion with sublimation in the freeze-etching technique (see Chap. 2.6), we shall use the term evaporation throughout this book] in a vacuum coating unit. A resistance evaporator is usually employed for this purpose. Carbon or graphite rods are sharpened to points which are then blunted to a diameter of about 1 mm. The rods are then mounted with the flat points touching one another and are held together under

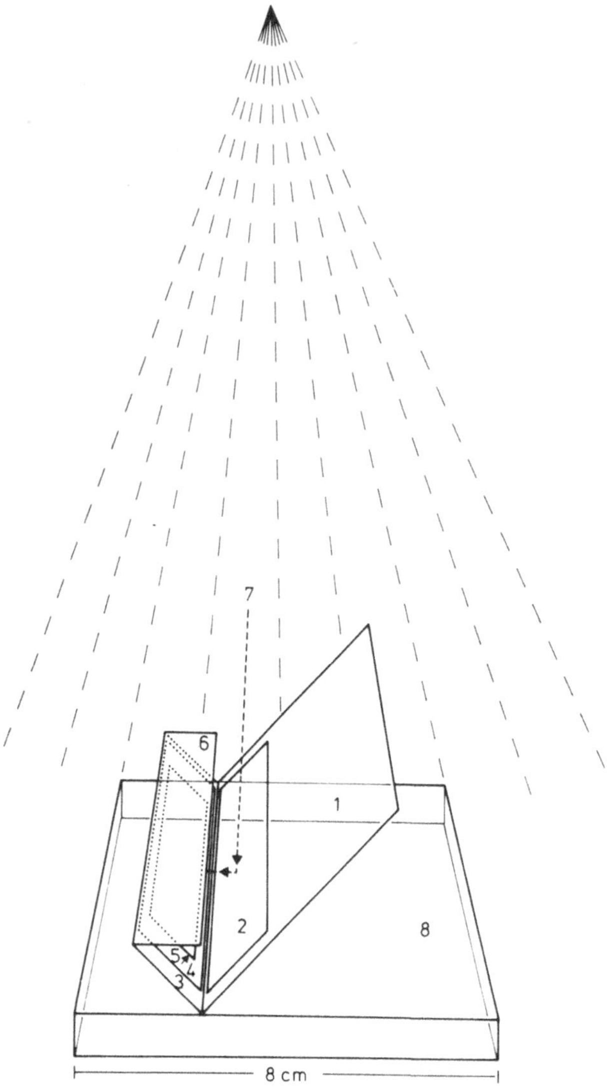

Fig. 1.10. Apparatus for indirect carbon evaporation. The beam of evaporated carbon particles *(7)* hits a microscope slide *(2)* which lies on a surface *(1)* tilted at 45° to the direction of the carbon beam. The carbon particles are reflected *(arrows)* onto a piece of cleaved mica *(5)* which lies on a second microscope slide *(3)* also tilted at 45°. A cover *(6)* shields the mica from the direct deposition of carbon. The various components are fixed to a heavy ground plate *(8)*

spring pressure (see also Fig. 2.43). The evaporation takes place through contact between the two carbon electrodes when an electric current is passed through them. A carbon film can be deposited on almost every type of surface; its thickness is proportional to the duration of the evaporation. Usually, however, carbon is deposited on freshly cleaved mica from which the film can easily be floated off onto water.

An indirect evaporation can be carried out when a particularly homogeneous carbon film is required. The equipment necessary for this is shown in Fig. 1.10. Here the evaporated carbon is reflected from the surface of a glass microscope slide set at an angle to the electrodes. The reflected carbon particles are deposited onto a neighbouring piece of cleaved mica which is shielded from direct deposition. For the evaporation the vacuum should be between 10^{-4} and 10^{-5} torr.

One can judge the thickness of the carbon film by eye if a film thickness monitor (see Chap. 2.6.5.3) is not available. A 2-nm-thick carbon film is hardly visible when floating on a water surface; in contrast a 10-nm-thick film is easily recognizable. Good films usually lie between 2 nm and 10 nm in thickness. Reproducibility in film thickness from one evaporation to another can easily be monitored by placing a piece of white card beside the slide. The intensity of the black colouration of the card gives a useful indication of film thickness.

If formvar or collodion filmed grids are to be coated with carbon they can be attached to the edge of a double-sided sticky tape (shiny side down) or inserted into a special holder which is commercially available. The filmed grids are positioned at 90° to the electrodes in the case of direct coating or can be mounted in a holder (Fig. 1.10) for indirect coating as described above. The latter method is to be recommended when the support films are very thin, since the deposition of the carbon layer occurs without a heat transfer.

Literature

Kölbel HK (1976) Kohleträgerfilme für die hochauflösende Elektronenmikroskopie. Verbesserung von Eigenschaften und Herstellungstechnik. Mikroskopie 32: 1-16

Lünsdorf H, Spiess E (1987) A rapid method of preparing perforated supporting foils for the thin carbon films used in high resolution transmission electron microscopy. J Microsc (in press)

Spiess E, Mayer F (1976) Herstellung einer Kohleträgerfolie (Film 1190) Wissenschaftlicher Lehrfilm des IWF Göttingen (Institut für den wissenschaftlichen Film, Nonnenstieg, D-3400 Göttingen). Aus der Reihe Elektronenmikroskopische Präparationsmethoden (also available in English)

2 Methods for TEM

2.1 Fixation, Dehydration and Embedding

2.1.1 Chemical Fixations

2.1.1.1 General Comments

Chemical fixation is probably the most frequently performed procedure in biological electron microscopy. With few exceptions (see Sect. 2.1.1.4 and 2.1.1.6) special pieces of equipment are not necessary: all that is needed is a bench centrifuge and normal laboratory glassware. The fixatives which are in general use are both poisonous and volatile, which means that (1) solutions should not be mouth-pipetted, (2) one works, whenever possible, in a fume cupboard, (3) the various containers for stock and waste solutions etc. should only be opened for short periods, (4) all glassware should be rinsed immediately after use, and (5) plastic gloves should be worn. Distilled H_2O should be used for rinsing the glassware both before and after contact with fixatives and all solutions should be made up with double-distilled H_2O. It is recommended that all fixatives and fixation solutions be stored in a refrigerator reserved especially for this purpose.

There are numerous books and articles in which fixations are described. Sometimes this is done in great detail, whereas in other cases only "cookery-book recipes" are given. Examples of both types are cited below.

Literature

Glauert AM (1975) Fixation, dehydration and embedding of biological specimens. In: Glauert AM (ed) Practical methods in electron microscopy, vol III, part 1. Elsevier/North Holland, Amsterdam, pp 1-207

Hayat M (1970) Principles and techniques of electron microscopy, vol I. Biological applications. Van Nostrand-Reinhold, New York, pp 1-412

Millonig G (1976) Laboratory manual of biological electron microscopy. Mario Saviolo, Vercelli, pp 1-67

2.1.1.2 Fixatives: Properties and Preparation

2.1.1.2.1 Osmium Tetroxide (OsO_4). OsO_4 is the most frequently used fixative. It is seldom used alone, instead it is applied either simultaneously with or after a glutaraldehyde fixation. When reduced, OsO_4 acts as an electron-dense stain which reacts principally with lipids. Stable glycol osmates are formed from conjugated double bonds, enabling cross-linking between neighbouring hydrocarbon chains to take place. This does not, however, explain the typical black-white-black trilaminar appearance of an osmium-fixed biomembrane. The latter apparently results from a displacement of the reduced osmium from the extremely hydrophobic environment of the reaction site towards the more polar hydrophilic portion of the lipid bilayer.

OsO_4 is available as a solution or as crystals in sealed ampoules. The latter form is to be preferred because of its greater storage life. The label on the ampoule is removed with either acetone or ethanol and then the ampoule is scored with a glass knife. It is then wrapped (using plastic gloves) in aluminium foil and broken open. The crystals, together with the glass, are carefully transferred to a glass-stoppered bottle. Water is added to give a final concentration of 2 or 4% (w/v) OsO_4. Because of the poor solubility of OsO_4 (maximum solubility is 7%, w/v, at 25° C) OsO_4 solutions are relatively stable at 4° C; they can develop a brown colouration due to the reduction of OsO_4 to metallic osmium (if this occurs the solution is unusable), and have a pH of 7. In comparison to aldehyde fixatives, those containing OsO_4 have a negligible osmolarity. However, cells fixed with OsO_4 are no longer osmotically active; their membranes lose their differential permeability properties.

2.1.1.2.2 Aldehydes. Of the various aldehydes which are employed as fixatives in EM glutar(di)aldehyde ($C_5H_8O_2$) is the most important. Acrolein (C_3H_5O) and paraformaldehyde (CH_2O) are also in use, particularly when a rapid penetration of the fixative is required. They all stabilize proteins by creating inter- and intrachain cross links. As a result of aldol condensations, several glutaraldehyde molecules can be linked together between neighbouring amino acid chains. Since aldehydes do not react with lipids, a subsequent fixation with OsO_4 is necessary to prevent losses occurring during dehydration.

Glutaraldehyde is normally delivered as a 25 or 50% solution. Sometimes it is contaminated with small amounts of polymeric aggregates which can be removed by shaking with charcoal (1 g: 5 ml solution. Check supernatant photometrically - contaminants absorb at 235 nm, glutaraldehyde at 280 nm). Stock solutions of 4 or 6% (v/v) can be prepared by simply diluting with water or buffer. These diluted solutions are only stable for several days at 4° C. Aqueous stock solutions of glutaraldehyde have a pH of 4-5. If the pH drops to below 3.5 the solution should be discarded.

Glutaraldehyde is osmotically active (a 3%, v/v, solution has an osmolarity of 300 mOsm) and shrinkages often result through its usage. However, biomembranes retain their permeability properties, and some enzymes still remain active in its presence. The use of glutaraldehyde as a primary fixative is always to be recommended when one is interested in preserving cytoplasmic (both cortical as well as spindle apparatus) microtubules.

Acrolein is available as a 100% (v/v) solution and can be stored for several months at 4° C. Usually it is used in conjunction with other aldehydes, but, because of its tear-gas-like properties, it should only be used as a last resort, i.e. when it becomes impossible to achieve better results with other aldehydes or when a rapid tissue penetration is of paramount importance.

Commercially available formaldehyde contains formic acid and methanol and is therefore unsuitable for EM. Instead, one usually prepares a solution of formaldehyde from paraformaldehyde. The latter is a powder, and an aqueous solution (40%, w/v) can be made by warming for 1 h at 65° C (constant stirring!). "Milkyness" can be overcome by adding a few drops of a 40% (w/v) NaOH solution. Formaldehyde solutions of this type are stable for several weeks at 4° C (in the dark). This fixative penetrates more quickly than glutaraldehyde but it fixes more slowly; moreover its action is reversible through washing. Formaldehyde is often used in combination with glutaraldehyde. In such cases the concentration of the formaldehyde is usually less than that of the glutaraldehyde.

2.1.1.2.3 Permanganate. Potassium permanganate ($KMnO_4$) was introduced as a fixative for EM by Luft (1956). It was used quite frequently, especially by botanists, in the years before fixing with aldehydes became the rule. Today it is seldom used because only membranes are preserved with it (Fig. 2.1 a–c). Furthermore, many membranous organelles, particularly mitochondria and plastids, swell as a result of a permanganate fixation. Ribosomes are not visible at all and chromatin is only poorly preserved in permanganate-fixed cells. This is because RNA- and DNA-containing structures are poorly fixed and are to a great extent lost during the subsequent dehydration step. Enzyme histochemistry is thus not possible with permanganate-fixed material. $KMnO_4$ is a strong oxidant and can be reduced to manganous dioxide (MnO_2). Precipitates of this electron-dense product become deposited in membranes, presumably through an interaction with the various hydrophilic groups present there. This leads to such an intense contrast in the EM that a post-staining of sections is often not necessary.

$KMnO_4$ dissolves slowly in water; several hours are necessary for the preparation of a 4% (w/v) stock solution. Such solutions do not retain their stability, even at 4° C, and should always be made up fresh for each new fixation.

Literature

Bradbury S, Meek GA (1960) A study of potassium permanganate "fixation" for electron microscopy. Q J Microsc Soc 101: 241–250

Luft JH (1956) Permanganate – a new fixative for electron microscopy. J Biophys Biochem Cytol 2: 799–802

Palade GE (1952) A study of fixation for electron microscopy. J Exp Med 95: 285

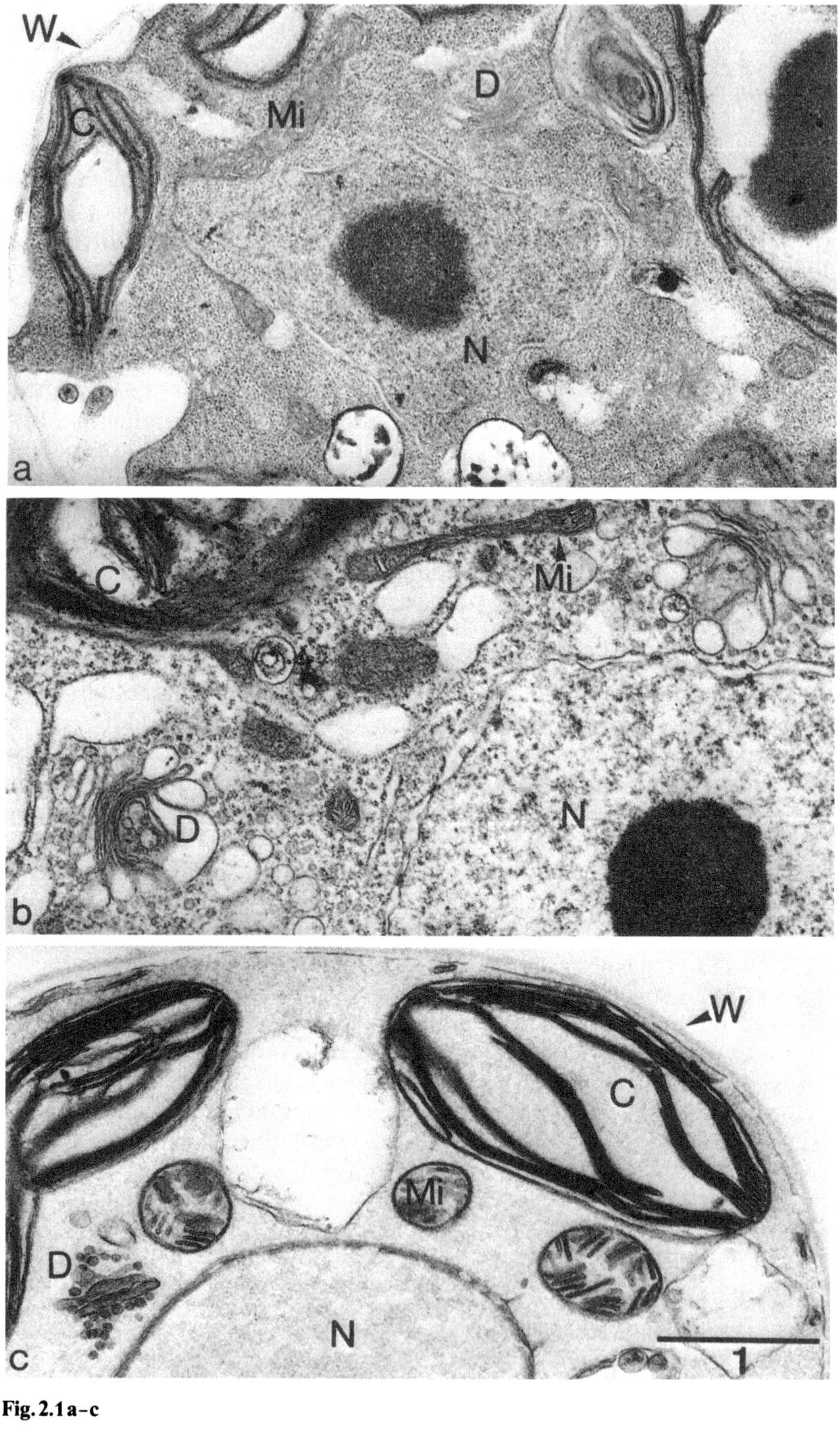

Fig. 2.1 a–c

Richards FM, Knowles JR (1968) Glutaraldehyde as a protein crosslinking reagent. J Mol Biol 37: 231-233
Riemersma JC (1968) Osmium tetroxide fixation of lipids for electron microscopy: a possible reaction mechanism. Biochim Biophys Acta 152: 718-727
Sabatini DD, Bensch K, Barnett RJ (1963) Cytochemistry and electron microscopy. The preservation of cellular structure and enzymatic activity by aldehyde fixation. J Cell Biol 17: 19-58

2.1.1.3 Composition of Fixation Solutions

With the exception of $KMnO_4$, which can be given to cells in H_2O, all other fixations are carried out in buffered solutions. In choosing the appropriate fixation "vehicle" three factors have to be considered: the range and capacity of the buffer, the ions present and the tonicity of the solution. In addition there are often other substances present which can modify the way in which the fixations work.

2.1.1.3.1 Properties of Buffers. There are numerous buffers which are in use in EM. The most frequently used are phosphate (Na/K), cacodylate and the various sulfonic acid derivatives, e.g. PIPES or MES. The reader is referred to Appendix 1 for details on the preparation of these solutions.

Usually fixation solutions have a pH of between 6.5 and 8.0; most are around pH 7. Since the pKa of buffers varies with temperature it is important to prepare the buffer solution at the temperature at which the fixation is to take place. The capacity of a buffer to maintain a given pH depends upon its pKa value and its concentration. As a rule the greater the pKa value differs from the desired pH, the more concentrated the buffer has to be in order to maintain the pH. Phosphate (pKa = 7.2) is a poor buffer for pH values above 7.5 and Tris (pKa = 8.3) is poor for pH values below 7.5. Unbuffered aldehyde solutions cause only small changes in intracellular pH; in contrast, unbuffered OsO_4 solutions give rise to pH value changes of between 2 and 3. Thus aldehyde fixatives do not have to be buffered as strongly as those containing OsO_4.

2.1.1.3.2 Ions. The presence of certain ions can be of great importance, particularly in OsO_4 fixations. Ca^{2+} and Mg^{2+} (1-3 mM) are especially useful in reducing fixative-induced swelling. In addition these cations contribute to the retention of a number of membrane components during dehydration.

Fig. 2.1 a-c. A comparison of three types of fixation as demonstrated on *Chlamydomonas reinhardii.* **a** Double fixation with 2% (w/v) glutaraldehyde in 0.1 M phosphate buffer for 1 h followed by 2% OsO_4 (w/v) in the same buffer, also for 1 h. **b** Simultaneous fixation with glutaraldehyde (1%, w/v, 0.1 M phosphate buffer at pH 7) given in equal volumes together at 4° C for 1 h, followed by an OsO_4 (2%, w/v, 0.1 M phosphate buffer) fixation at 4° C for 3 h. **c** Permanganate fixation with 1% $KMnO_4$ in H_2O at 4° C for 2 h. Sections **a** and **b** were post-stained with uranyl acetate and lead citrate; section **c** stained with lead citrate alone. *C* chloroplast; *D* dictyosome; *Mi* mitochondria; *N* nucleus; *W* cell wall. Magnification bar in µm

2.1.1.3.3 Tonicity. Cells can swell or shrink depending on the osmotic capabilities of a fixation solution. In order to avoid such artefacts, careful attention has to be paid to the concentration and composition of a fixation solution. In theory one tries to attain isotonic conditions during the fixation, but in practice, especially with animal cells, one fixes somewhat hypertonically. Of particular importance is the tonicity of the primary fixation solution (usually an aldehyde). Although the aldehydes have an appreciable osmolarity, this is often ignored when preparing fixation solutions for many animal objects (see also Sect. 2.1.1.4.2). Instead, the osmolarity is usually adjusted through the fixation vehicle. The osmolarity (given as Osmol l^{-1}) and the osmolality (given as Osmol kg^{-1}) of fixation solutions are easily determined with the help of an osmometer on the basis of freezing point-depression measurements.

Table 2.1 presents the osmolarity of a number of standard buffers and fixatives. Minor adjustments in the osmolarity of fixation solutions can be achieved through the addition of sugars (glucose, sucrose) or cations (K^+, Na^+, Mg^{2+}, Ca^{2+}). As a rule NaCl is used. It has been claimed that the presence of sugars can lead to leaching but there is often no evidence of this in published micrographs. Moreover sugars do not influence the pH of the fixation solution. Finally, it is important that the osmolarity of the wash/rinse solutions and of the secondary fixation solution is the same as that of the primary fixation solution.

Table 2.1. Osmolarity of some buffer and fixation solutions

Concentration	Substance	Osmolarity (mOsmol l^{-1})
2.5%[a]	Glutaraldehyde in H_2O	270
2%	Formaldehyde in H_2O	215
0.05%	Phosphate buffer pH 7.4	145
0.1 M	Phosphate buffer pH 7.4	240
0.2 M	Phosphate buffer pH 7.4	420
0.05 M	Sodium cacodylate buffer pH 7.4	110
0.1 M	Sodium cacodylate buffer pH 7.4	195
0.2 M	Sodium cacodylate buffer pH 7.4	390
0.05 M	PIPES buffer pH 7.4	120
0.1 M	PIPES buffer pH 7.4	236
0.2 M	PIPES buffer pH 7.4	325
2.5%	Glutaraldehyde in 0.1 M phosphate pH 7.4	480
2.5%	Glutaraldehyde plus 2% formaldehyde in 0.05 M cacodylate pH 7.4	550
2.5%	Glutaraldehyde plus 2% formaldehyde in 0.1 M cacodylate pH 7.4	900
2.5%	Glutaraldehyde in 0.1 M PIPES pH 7.4	345
2.5%	Glutaraldehyde plus 4%[b] NaCl in 0.1 M phosphate pH 7.4	1600
2.5%	Glutaraldehyde plus 10%[b] sucrose in 0.1 M phosphate pH 7.4	800

[a] % values given as v/v.
[b] % values given as w/v.

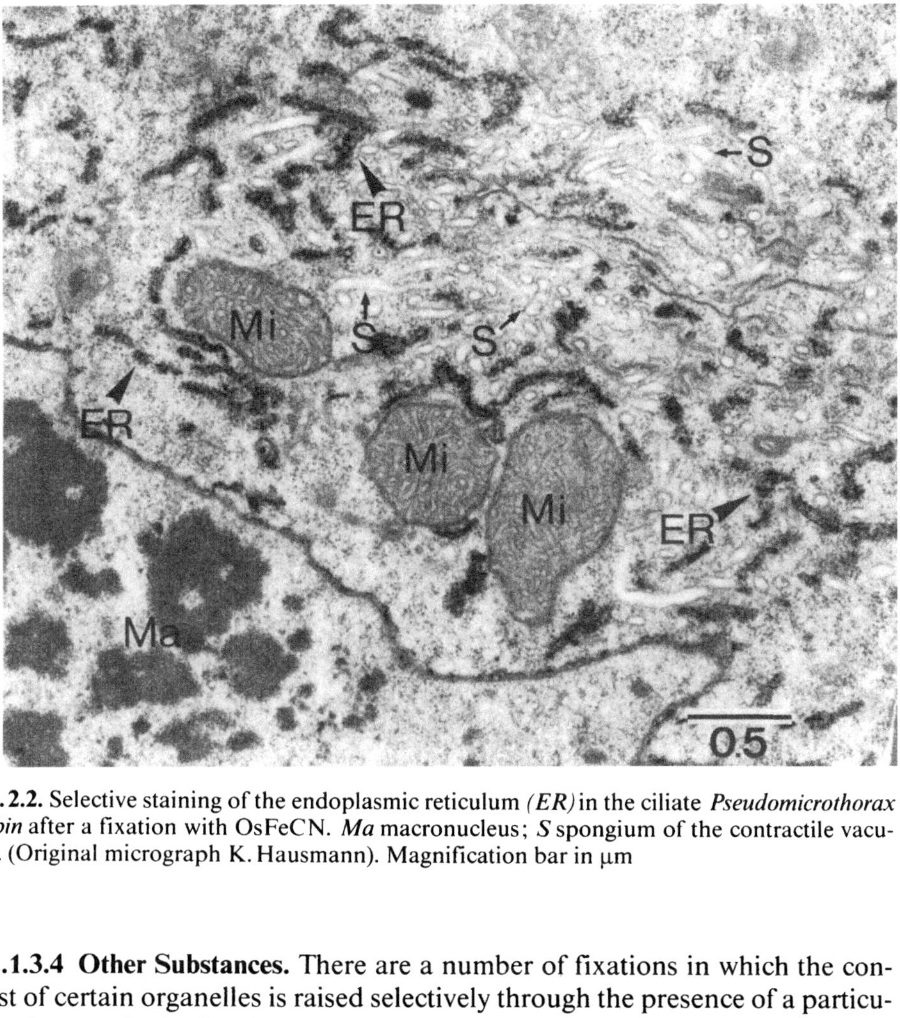

Fig. 2.2. Selective staining of the endoplasmic reticulum *(ER)* in the ciliate *Pseudomicrothorax dubin* after a fixation with OsFeCN. *Ma* macronucleus; *S* spongium of the contractile vacuole. (Original micrograph K. Hausmann). Magnification bar in μm

2.1.1.3.4 Other Substances. There are a number of fixations in which the contrast of certain organelles is raised selectively through the presence of a particular substance in the fixation solution. The three best-known examples of this are potassium dichromate, potassium ferricyanide and tannic acid.

Potassium dichromate (together with KOH) was introduced by Dalton (1955) as a strong buffer for OsO_4 fixations. Its use, however, led to a stronger staining of the cytoplasm. Although in the meantime other buffer systems have proved more effective, the inclusion of potassium ferricyanide (0.5–1%, w/v) in the OsO_4 solutions (1–2%, w/v) leads to an enhanced staining of membranes. Since the reaction is improved in the presence of Ca^{2+} ions there is normally 1–10 mM Ca^{2+} present in both the primary fixation and wash/rinse solutions. In many, but not all, cases, it is the ER which appears to be selectively stained (Fig. 2.2). It has been argued that the reduction of OsO_4 through ferricyanide is related to the existence of an electron transport system in the ER.

Tannic acid is often included in fixation solutions in order to improve the visualization of fibrillar elements [e.g. microtubules (see Fig. 2.3 a, b) and micro-

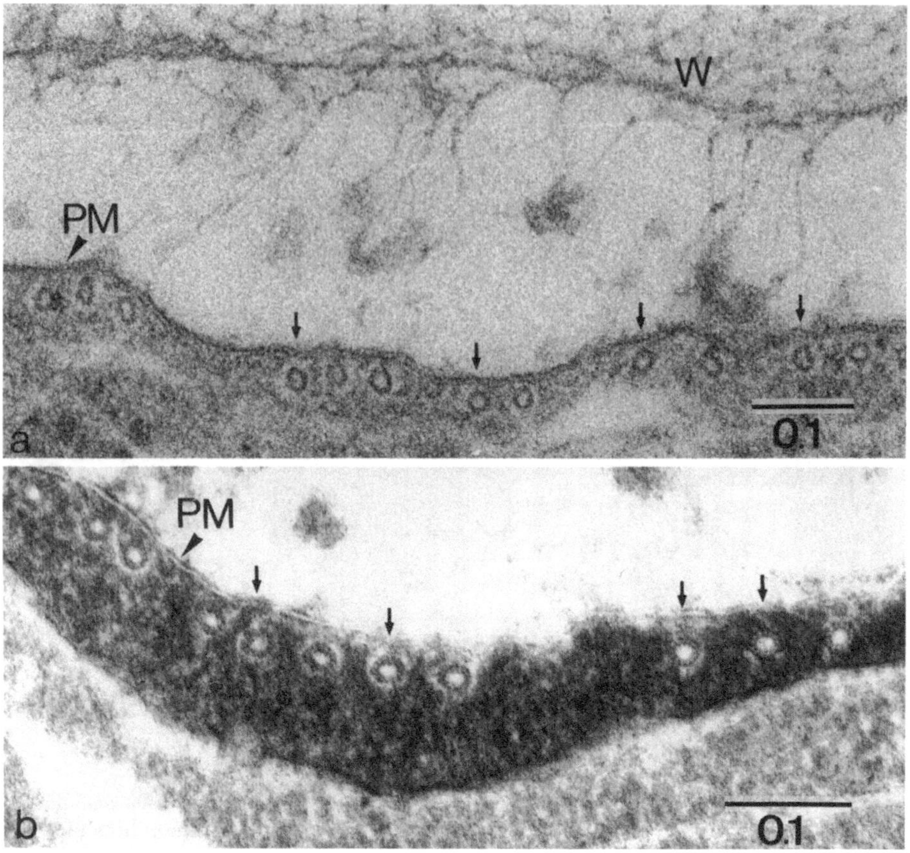

Fig. 2.3 a, b. Improved visualization of cortical microtubules *(arrows)* in the seed hairs of *Cobaea scandens* as a result of fixing in the presence of tannic acid. **a** A conventional double fixation. **b** A fixation with tannic acid in the primary fixation solution. *W* cell wall; *PM* plasma membrane. (Original micrographs E. Schnepf). Magnification bar in μm

filaments] and the plasma membrane (e.g. the clathrin framework around "coated pits"). Although tannic acid is known to act as a mordant and causes OsO_4-stained structures to stain more intensely with lead salt solutions, the molecular basis of this phenomenon is not known. Usually the primary fixation solution contains between 1 and 8% (w/v) tannic acid. Sometimes 0.05% (w/v) saponin is added to the primary fixation solution in order to increase the permeability of the cells towards tannic acid. Tannic acid has also been presented successfully to cells after the secondary fixation with OsO_4. In this case the tannic acid is simply dissolved in buffer solution and given to the cells at room temperature. The cells are then rinsed quickly (5 min) in a 1% (w/v) Na_2SO_4 solution before dehydrating.

Literature

Bone Q, Ryan KP (1972) Osmolarity of osmium tetroxide and glutaraldehyde fixatives. Histochem J 4: 331-347
Dalton AJ (1955) A chrome-osmium fixative for electron microscopy. Anat Rec 121: 28
Gomori G (1955) Preparation of buffers for use in enzyme studies. In: Clowich HSP, Kaplan NO (eds) Methods in enzymology, vol I. Academic Press, London New York, pp 138-146
Good NE, Winget GD, Winter W, Conolly LTN, Zawa SI, Singh RMM (1966) Hydrogen ion buffers for biological research. Biochemistry 5: 467-477
Maser MD, Powell TE, Philpott CW (1967) Relationships among pH, osmolality and concentration of fixative solutions. Stain Technol 42: 175-182
Maupin P, Pollard TD (1983) Improved preservation and staining of HeLa cell actin filaments, clathrin-coated membranes, and other cytoplasmic structures by tannic acid-glutaraldehyde-saponin fixation. J Cell Biol 96: 51-62
Simionescu N, Simionescu M (1976) Galloylglucoses of low molecular weight as mordant in electron microscopy. I. Procedure, and evidence for mordanting effect. J Cell Biol 70: 608-621
White DL, Mazurkiewiez JE, Barnett RJ (1979) A chemical mechanism for tissue staining by osmium tetroxide-ferricyanide mixtures. J Histochem 27: 1084-1094

2.1.1.4 The Fixation of Animal Cells

There is no patent recipe for the successful preservation of structure in animal cells. Although suitable fixations can only be achieved on a trial and error basis, particular attention will be given here to the fixation procedure and to the choice of a particular fixation solution.

2.1.1.4.1 Fixation Procedures. There are two important problems often to be considered when fixing animal cells or organisms:

1. the death of parts of the organism through O_2 deficit (anoxia) before the fixation has been completed, and
2. mechanical damage before or after the fixation.

It is, for example, well known that representatives of the higher vertebrates are especially sensitive to O_2 deficit and that autolysis can occur within minutes after cell death. Structural and enzymatic changes occur as a result of this. Depending on the object in question, there are four ways to fix animal organisms.

Immersion Fixation

Here whole objects or parts of them are placed directly in the fixation solution. The primary fixative is usually an aldehyde (2 h at 4° C) and the secondary fixative is normally OsO_4 (2 h at 4° C). This procedure is particularly suitable for small invertebrates because of the small diffusion pathways involved. A prerequisite for the good fixation of whole animals is the quick penetration of the external integument and the internal tissues. This is aided considerably when the fixation is carried out under mild vacuum conditions (exsiccator with a water suction pump).

With animals larger than 1 mm^3 (e.g. with annelids, crustaceans and insects) the desired tissue has to be laid free before the fixation can be started. Dissection times of up to 30 min at room temperature on anaesthetized animals can very often be tolerated without the loss of structural integrity. In some cases the dissection must be carried out in a cold laboratory (4° – 10° C). This temperature leads to an effective immobilisation of cold-blooded (poikothermic) animals.

During the dissection care must be taken to avoid desiccation. This can easily be done by applying a physiologically balanced salt solution during the dissection but fixation during dissection is to be preferred. Care must also be taken to avoid mechanical distortion or damage to the organ in question. This is, with some organs, e.g. with embryonic tissue, the nervous system, and organs of perception, almost unavoidable. In such cases the fixation solution (usually an aldehyde) should be added dropwise to the organ in question as soon as it becomes exposed. This procedure should be continued until the organ is fully free and can be placed in the primary fixation solution. As a rule the duration of the dissection, when aldehydes are allowed to percolate onto the object, should not exceed 10 min.

Perfusion Fixation

When organs are too large or dissecting is difficult, fixation is carried out by perfusing the intact organism with fixation solution. This is achieved by injecting the fixative either into the vascular system or into the interstitial spaces. The latter method is best carried out on small invertebrates. For those animals with a hard exoskeleton (insects, crustaceans) the injection should be made through the softer portions of the integument, i.e. between the segments. Normally a distally situated extremity, e.g. the head, when the injection is made into the abdomen, must be opened before the injection. This allows body fluids and excess fixation solution to leave the organism without causing internal pressure changes. Only through a careful and continuous injection of the fixation solution, as indicated by the dropwise displacement of body fluids at the severed portion, can one avoid the deformation or drifting of internal organs.

For many vertebrates, and in particular for organs which are difficult to dissect, e.g. the nervous system, a vascular perfusion is the method of choice. The method differs according to the animal in question, being in general more complicated for large warm-blooded animals than for small, cold-blooded representatives. Thus for mammals artificial respiration via the trachea is often necessary, which is usually not so for amphibians, birds and some smaller mammals, providing the experimenter works quickly. The actual site of the infusion can vary and depends on the tissue to be fixed and the size of the animal. In general, however, one tries to perfuse directly into the heart or the aorta.

We give below a description of a heart perfusion as it can be performed on a frog, and refer the reader to the detailed account of perfusion fixation in Hayat (1981) for additional information.

Preparing for a Perfusion Fixation

Instruments

The following are necessary: a plastic basin or developing tray with a fitting cork or wax plate upon which the animal can be laid out and pinned-down; pins and a dissecting kit (see Fig. 2.4). For larger animals (and for certain purposes, e.g. removal of the brain from the skull) large bone shears can be of considerable use.

Perfusion Devices

Pressure determines the speed of perfusion. This should be sufficiently high to displace the blood but not much higher than the natural arterial pressure of the living animal. To this end one can use commercial pumps or a dropping device as in blood transfusions. In the latter case plastic infusion bags containing the primary fixation solution and wash/rinse solutions are suspended above the animal at a height of about 50–150 cm. The flow rate is controlled with clamps and the choice of solution can be determined with a threeway tap (see Fig. 2.5). Care must be taken to avoid air bubbles entering the system. As an alternative large disposable plastic syringes can be used instead of the gravity-dropping device just described. If the volume of the pipette is insufficient and a new pipette has to be attached, great care must be taken to avoid damaging the point of en-

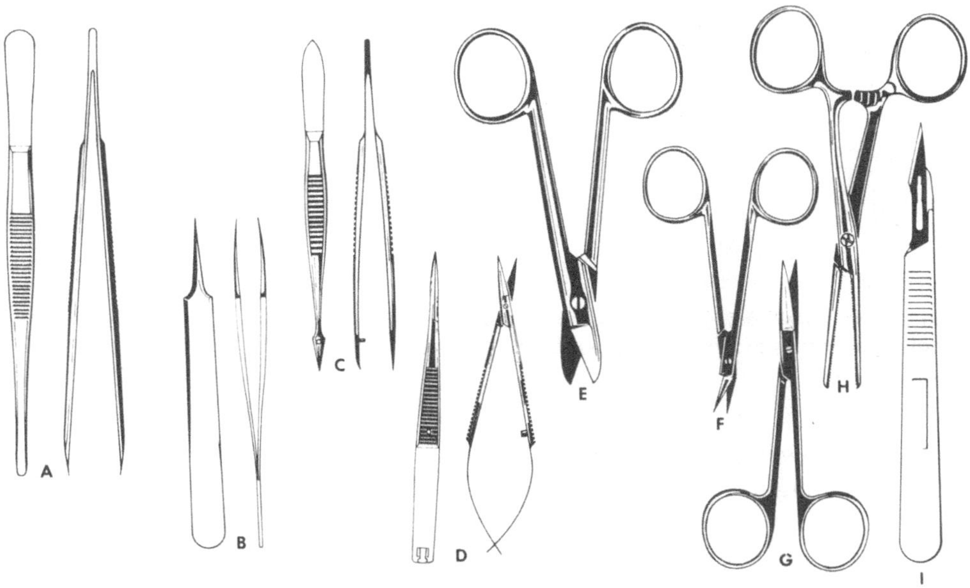

Fig. 2.4 A–I. Instruments needed for dissection: **A** dissecting forceps; **B** watchmaker's forceps; **C** microscopic forceps; **D** isidectomy scissors; **E** scissors of physiology; **F** angular microdissecting scissors; **G** microdissecting scissors; **H** haemostatic forceps; **I** scalpel

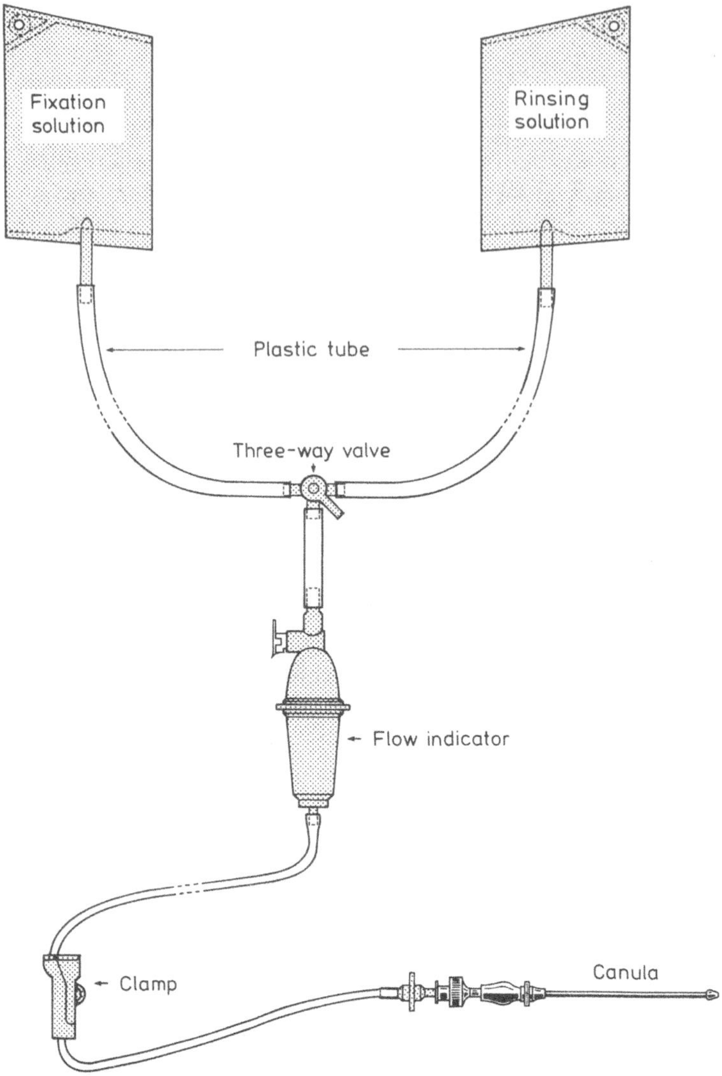

Fig. 2.5. A simple device for perfusion fixation. Pre-rinse and primary fixative solutions are contained in plastic infusion bags which are hung on a clamp stand

try of the injection needle. Typical perfusion volumes are 100 ml for a frog and 250 ml for a rat.

Solutions

Perfusion fixations are, as a rule, only successful when the blood is rinsed out of the vascular system prior to administration of the fixative. This is necessary because the fixative can lead to the aggregation of blood cells, causing in turn a blockage of the vascular system. Usually a buffer or a commercially available blood substitute is passed through the vascular system before the primary fixative is applied. These pre-rinse solutions often contain dilating substances, e.g. procaine (0.5%, w/v), anticoagulants, e.g. heparin (0.25%, w/v) and substances which are known to raise the colloidal osmotic pressure, e.g. dextran or polyvinylpyrrolidone (2.5%, w/v). The effect of the latter agents is to prevent the enlargement of extra vascular spaces. In the case of a frog or a rat the primary fixation solution usually consists of 2.5% glutaraldehyde (or a mixture of 1% formaldehyde and 2% glutaraldehyde) in 0.1 M phosphate or cacodylate buffers at pH 7.4 (for buffers see Appendix 1). The formaldehyde is freshly prepared from paraformaldehyde (see Sect. 2.1.1.2.2). The secondary fixation is with OsO_4 (usually 2%; w/v) in the same buffer system. Both fixations are carried out with pre-cooled (4° C) solutions and all solutions contain 1 mM $CaCl_2$ (see Sect. 2.1.1.3).

Performing a Perfusion Fixation

The animal must be well anaesthetized before opening the thorax. For small animals, e.g. frogs, this can be achieved with ether; better, and for larger animals to be recommended, is an injection with nembutal or urethane (ethyl carbamate) either subcutaneously, intravenously or into the peritoneum (for dosage see Hayat 1981).

The following description of an intracardiac perfusion applies to a frog *(Xenopus):*

1. After narcotising with ethyl carbamate (10%, w/v, 0.8 ml) the animal is pinned down on its back.

2. The skin is cut open from the tip of the breast bone to the jaw. The thorax is opened by a midline ventral incision through the body wall and pectoral girdle, close to bone and cartilage. The sternal bone (xiphisternum) is cut by a horizontal cut so that a window-like portion of tissue can be removed. The whitish pericardium is carefully raised with a pair of blunt forceps and cut open with small pointed scissors. The heart now lies exposed with the bulbus cordis at the left leading into the truncus arteriosus (arterial arches, see Fig. 2.6). With the help of a pair of pointed tweezers one raises the truncus and pulls a piece of fine thread behind it. A loosely fitting loop is then made around the truncus. An incision in the lower right-hand portion of the bulbus is made with a scalpel and the injection needle pushed upward into the truncus. The thread is then tightened around the injection needle. It is important that the needle is not inserted too far into the truncus; both of the major arteries leaving the truncus must remain free. In order to assist the drainage of blood, one makes an incision into the sinus venosus on the dorsal side of the heart (the heart has to be raised somewhat when the animal is lying on its back).

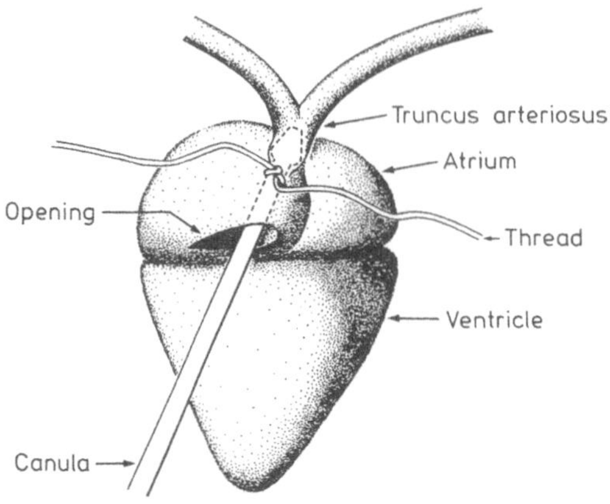

Fig. 2.6. Drawing of a frog heart with inserted perfusion canula (ventral view)

3. The pre-rinse solution is then allowed to flow into the truncus, until all of the blood has run out of the sinus venosus (takes about 1 min and is recognized by the lighter colour of the effluent liquid).

4. The pre-rinse solution is shut off, and the primary fixative solution allowed to flow for about 5 min. A successful fixation can be judged by the loss of trembling movements at the extremities and by the hardening of organs and stiffening of limbs. Further proof of a good primary fixation is the lighter colour of the liver and its hardness.

5. After the perfusion has been completed, the partially opened, fixed animal is then immersed in the primary fixative for at least 2 h at 4° C. During this period the various organs can be dissected out as desired. After being washed in buffer, the organ in question is placed in secondary fixative solution.

Unfortunately, perfusion fixations, even in the hands of the most experienced of workers, do not always lead to uniformly good results; some organs show regions with good structural preservation adjacent to others where this is not the case.

Fixation by Injection

Sometimes primary fixation solutions are injected directly into a particular organ or tissue. Afterwards the object is cut out and cut into smaller pieces for the secondary fixation. This somewhat harsh method often leads to quite reasonable results. For compact tissues such as the liver or ganglia the tissue blocks should not exceed 1 mm^3, enabling a good penetration with OsO_4. Often, if not always, one tries to prepare the tissue blocks with a specific orientation, e.g. to the main axis of the body. In cutting the tissue (with a new, sharp razor blade) it is important not to apply too much pressure – the razor blade should be pulled gently through the tissue. The arm of a pair of fine tweezers can be used as a guide for this cutting action, which should be carried out in liquid under a binocular microscope. For particularly sensitive material the tissue can be embedded in agarose (3%, w/v, warmed to 45° C then allowed to cool after insertion of the tissue) and then cut into blocks (see also Andres and von Düring 1981). As a rule the tissue blocks are washed/rinsed in buffer (same composition as for the primary fixation solution) several times over a 5-min period and then fixed in a buffered OsO_4 solution for 2 h.

2.1.1.4.2 Choice and Composition of a Fixation Solution. Animal tissues are usually double-fixed, the primary aldehyde fixation (2 h at 4° C) being followed by one with OsO_4 (also 2 h at 4° C). Single fixations with OsO_4 alone or simultaneous fixations with an aldehyde and OsO_4 are much less in use. Triple fixations (with uranyl acetate following the osmification) are quite often carried out and are to be recommended when details of membrane structure are desired (see also Sect. 2.1.1.2).

Formaldehyde concentrations of between 1 and 4% (w/v) and glutaraldehyde concentrations of 1-6% (v/v) can be found in the literature; usually, however, aldehyde concentrations of 1-2.5% are sufficient. Lower aldehyde concentrations are to be recommended for sensitive tissues, e.g. embryos, marine animals and tissues containing a lot of water.

Body fluids, tissues and cells can have different osmotic values depending on the animal and its particular physiological condition. Since the osmotic pressure of the fixation solution can considerably influence structural integrity, it is important to appropriately adjust the osmolarity of the fixation solution. As a guide, the osmolarity of the fixation solution should correspond to that of the body fluid or blood, i.e. it should be iso- or slightly hypertonic. Solutions which are too hypertonic lead to shrinkage and deformation of the organ or tissue in question; solutions which are too hypotonic lead, in contrast, to a swelling. Values for the osmotic pressures of blood or body fluids in animal cells can be found in numerous books, e.g. Potts and Parry (1964).

The fixation of marine animals, especially when they are small and sensitive, is beset with difficulties. Unfortunately there is no easy formula for success. In general, the osmolarity of the fixation solution is higher (particularly for invertebrates) than for freshwater or land animals. This is normally achieved by the addition of NaCl (usually 4%, w/v) or sucrose (up to 10%, w/v). A good fixation schedule for invertebrates is: primary fixation for 1 h at room temperature with 2.5% glutaraldehyde + 0.14 M NaCl in 0.2 M phosphate buffer pH 7.4 followed by a secondary fixation, also for 1 h at room temperature, with 2% OsO_4 + 1.25% (w/v) sodium bicarbonate at pH 7.4 (pH adjusted with HCl). In this fixation it is important that the primary fixation solution be changed several times during the primary fixation, otherwise a precipitate of calcium phosphate will be formed from the calcium brought in with the seawater and the phosphate buffer. Between the primary and secondary fixation the object is washed several times with 0.2 M phosphate buffer containing 0.3 M NaCl.

A simultaneous fixation is often useful for small objects (e.g. embryos or invertebrates). Two stock solutions are necessary: solution A – contains

5% (v/v) glutaraldehyde and 3% NaCl; solution B – contains 4% OsO_4 and 3% NaCl.

The solutions are mixed together (1:1, v/v) immediately before use. Both solutions contain 0.2 M cacodylate buffer pH 7.4, and the final osmolarity of the fixative, obtained by mixing equal volumes of A and B, is 1060 mOsmol. (One alters the concentration of NaCl accordingly), should another osmolarity be required. The fixation is carried out at 4° C. The object is then washed (3 × 10 min) in 0.1 M cacodylate buffer before dehydrating and embedding. If desired, the object can be embedded in 2% agarose (see above) before dehydrating.

Literature

Andres KH, Düring M von (1981) General methods for characterization of brain regions. In: Heym C, Forssmann WG (eds) Techniques in neuroanatomical research. Springer, Berlin Heidelberg New York, pp 100-108

Cloney A, Florey L (1968) Ultrastructure of cephalopod chromatophore organs. Z Zellforsch 89: 250-280

Dorresteijn AWC, Bilinski SM, Biggelaar JAM van den, Bluemink JG (1982) The presence of gap junctions during early *Patella* embryogenesis: an electron microscopical study. Dev Biol 91: 397-401

Glauert AM (1975) Fixation, dehydration and embedding of biological specimens. In: Glauert AM (ed) Practical methods in electron microscopy, vol III, part 1. Elsevier/North Holland, Amsterdam, pp 1-207

Hayat MA (1981) Fixation for electron microscopy. Academic Press, London New York, pp 1-501

Potts WTW, Parry G (1964) Osmotic and ionic regulation in animals. Pergamon, Oxford, pp 1-423

2.1.1.5 The Fixation of Plants and Microorganisms

2.1.1.5.1 Plant Tissue. In comparison to animal tissues, the accessibility of cells in plant tissues to fixation solutions is not a major problem. Immersion fixation is the rule even if the tissue is dissected beforehand. The dissection can, however, occur in the presence of the fixative. Because of the air-filled intercellular spaces in plants it is advisable to carry out the first 15 min of the primary fixation in vacuo (i.e. in an exsiccator attached to a water suction pump).

For tissue blocks of 1 mm^3, 1-3 ml of fixation solution is sufficient. The primary fixation solution usually contains 2-6% (v/v) glutaraldehyde in 0.02-0.1 M buffer at pH 7. Fixations are normally carried out at room temperature and last 1-3 h, although periods of up to 24 h are not seldom for woody tissue. In the latter case the tissue blocks are usually transferred to the refrigerator (4° C) after 1-2 h of fixing at room temperature, in order to avoid cell autolysis.

After the primary fixation the blocks are washed in buffer solution (per object 2-3 ml; four changes) and then transferred to the secondary fixation solution. This is usually 1-2% (w/v) OsO_4 in the same buffer. Once again the fixation is for 1-2 h at room temperature or in the cold.

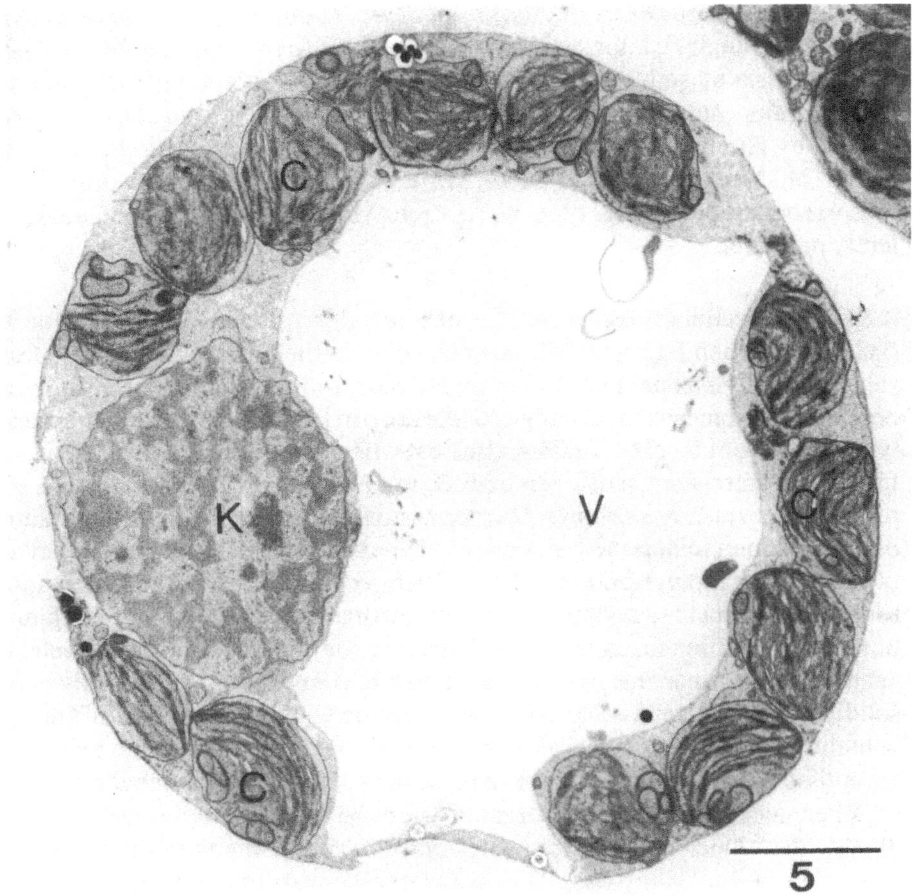

Fig. 2.7. Mesophyll protoplast from *Vicia faba* L. Single fixation with OsO_4 (2%, w/v, in 0.1 M phosphate buffer pH 7; 2 h in an ice bath). *C* chloroplast; *N* nucleus; *V* vacuole. Magnification bar in µm

2.1.1.5.2 Protoplasts. Protoplasts are prepared from plant tissues through the application of polysaccharidase enzyme mixtures. Since the digestion of the cell wall occurs under plasmolyzing conditions (presence of 0.4–0.8 M mannitol or sorbitol) the plasmolyzing agent must also be present in the primary fixation solution. This contains, as for normal tissue fixations, 1–2% (v/v) glutaraldehyde and 0.1–0.2 M buffer at pH 7. After fixing for 1–2 h at room temperature, the protoplasts are washed carefully in buffer containing the plasmolyzing agent, and then transferred to a solution of 1–2% (w/v) OsO_4 in 0.05 M buffer (pH 7) without the plasmolyzing agent. After 1–2 h at room or cold temperatures, the protoplasts can be dehydrated. Protoplasts are very sensitive, making it advisable not to centrifuge them during the fixation procedure. As a rule, however, protoplasts sediment without centrifugation so that solutions can easily be changed by pipetting off the supernatants.

Very good protoplast fixations (see for example Fig. 2.7) can also be achieved by single fixations with 0.1 M phosphate (pH 7) buffered OsO_4 (1-2%, w/v, *without* a plasmolyzing agent) or by the gradual addition of OsO_4 to a primary fixation solution (buffered aldehyde plus plasmolyzing agent) as described by Hillmer et al. 1986. These fixations are carried out in the cold and take 1-2 h. Microtubules and microfilaments tend not to be, or are only poorly, preserved with these latter types of fixation. Membranes, in contrast, are excellently preserved.

2.1.1.5.3 Unicellular Organisms. Unicellular algae, protozoa and bacteria are fixed by resuspending a centrifuged cell pellet in the fixative. Algae and protozoa are usually pelletable at 1500 *g* in 1-2 min; for bacteria higher *g* values are necessary. This means that changes of solution in the case of the former can easily be carried out by centrifugation but not when it comes to bacteria. Thus one finds that bacteria are usually embedded in agar after the primary fixation. The procedure for this is a follows: 0.1 g agar is dissolved in 10 ml boiling fixation buffer and then maintained in a water bath at 45° C. The cells are pelleted out of the primary fixative and the pellet is warmed to 45° C. Molten agar is added to the cell pellet (1:1, v/v), stirred rapidly with a vortex mixer and then poured out onto a cold microscope slide. If there are only small amounts of cells the agar-cell mixture can instead be sucked up into a Pasteur pipette. Shortly before solidification, the agar is blown out of the pipette. Small segments of the agar containing the embedded cells can be cut with a razor blade and are to be treated as tissue blocks for the subsequent secondary fixation and dehydration.

The majority of unicellular organisms can be fixed with a double fixation as already described, although for some algae the culture medium can be used instead of a buffer. Many flagellates show excellent structural preservation as a result of a simultaneous fixation with glutaraldehyde and osmium tetroxide (see Fig. 2.1 b, also Sect. 2.1.1.4.2). For this 2% (w/v) solutions of glutaraldehyde and OsO_4 in 0.1 M cacodylate or phosphate buffers at pH 7 are required. The solutions are mixed 1:1 (v/v) in an ice bath immediately before use. The fixation is carried out for 1 h in the cold. After washing (four changes in buffer, 1 h) the cells are transferred to pH 7 buffered OsO_4 (2%, v/v) at 4° C for 2-3 h. After being washed again in buffer, the cells can be dehydrated, or treated with uranyl acetate (see Sect. 2.2.6.5) before dehydration. One should note that simultaneous fixations of this type are often unsuitable for higher plants, since as a result the plasma membrane tends to rupture into small pieces.

Literature

Aldrich HC, Todd WJ (eds) (1986) Ultrastructure techniques for microorganisms. Plenum, New York, pp 1-545

Fowke LC (1975) Electron microscopy of protoplasts. In: Gamborg OL, Wetter LR (eds) Plant tissue culture methods. Nat Res Counc Can, Ottawa, pp 55-62

Franke WW, Krien S, Brown RM (1969) Simultaneous glutaraldehyde-osmium tetroxide fixation with postosmication. Histochemistry 19: 162-164
Hillmer S, Depta H, Robinson DG (1986) Confirmation of endocytosis in higher plant protoplasts using lectin-gold conjugates. Eur J Cell Biol 41: 142-149

2.1.1.6 The Fixation of Isolated Organelles

When fixing isolated organelles the problem of long diffusion pathways can be almost totally neglected. Thus the fixation periods are correspondingly much shorter than is the case with tissues or whole cells. Since the majority of organelles are isolated with the help of a density gradient it is expedient to fix in the presence of the gradient medium. Moreover, because gradient media are usually buffered one only needs to add the appropriate volume of 25% (w/v) glutaraldehyde so that the end concentration of the aldehyde is around 2% (w/v). After the primary fixation (usually at 4° C) the organelle suspension is diluted with homogenizing buffer and centrifuged at 100,000 *g* for 30-60 min. The resulting pellet is carefully washed with three to four changes of buffer over a 30-min period (the pellet should not be resuspended!). The secondary fixation is carried out by adding 0.5 ml of a 1-2% (w/v) OsO_4 solution (buffered with 0.1 M cacodylate or phosphate at pH 7). After about 30 min the pellet is hard enough to be broken up into small pieces which should be left in the fixative for a further 1-2 h.

With the method just described it is not possible to take into account heterogeneities in the organelle fraction which has been fixed. In order to do this it is necessary to pellet the organelles as a thin layer onto the surface of a 3 mm diameter Millipore filter disc (porosity: 0.025-0.4 μm according to the size of the organelles to be fixed). Since this layer is very sensitive, subsequent manipulations should be kept to a minimum. A useful procedure in this regard involves a step gradient and enables the primary fixation, the washing step and the secondary fixation to be accomplished in one centrifugation (see Fig. 2.8). The filter disc with the fixed organelles on it is then carefully removed from the centrifuge tube with fine tweezers and brought into a depression of a porcelain

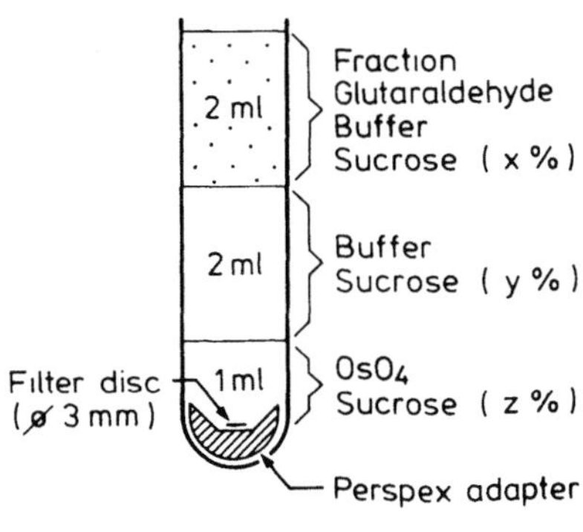

Fig. 2.8. Fixation of isolated organelles with the help of a step gradient. The steps contain increasing concentrations (z > y > x) whereby x is less than the isopycnic density of the organelles. The organelles are centrifuged at 100,000 *g* for 1 h (swing-out rotor) onto a Millipore filter disc which fits into a specially made perspex adapter

staining tray. Washing and dehydration, with a minimum of liquid, can be carried out in the staining tray. The millipore filter dissolves in the last acetone steps leaving a thin film (up to 50 μm thick) of fixed organelles. A flat embeddment (see Sect. 2.1.3.5) is to be recommended in order to section exactly perpendicular to this layer.

2.1.1.7 Fixing for Immunocytochemistry

The basic principles of immune biology as applied to EM are given in Section 2.4. Here we shall deal with a preparatory aspect of immunocytochemistry, but one which is nevertheless of great importance. Essentially, fixing for immunocytochemical purposes represents a compromise. On the one hand, cellular components have to be stabilized in order to be successfully embedded; on the other hand, antigenic properties have to be retained as much as possible. Unfortunately, there is no ready answer to this dilemma; as in many areas of biological electron microscopy it is a question of trial and error.

Usually the primary fixation is short (1 h) and is carried out in the cold, but the choice and concentration of the primary fixative (glutaraldehyde, paraformaldehyde) can often only be determined by testing the effects of these fixations on the antigen in advance. This can be done quite simply by a double immunodiffusion test or by Western blotting, both carried out in the presence of different concentrations of fixative.

When an antigen is to be localized with the help of cryosections a secondary fixation (with OsO_4) is not performed. Many workers also avoid exposing their specimens to OsO_4 when embedding and sectioning conventionally. If, however, for reasons of better structural preservation and contrast, cells have to be fixed with OsO_4, the masking effects of this fixative can be partially reversed by treating the sections with sodium metaperiodate (see Sect. 2.4.2.2). There is, however, no guarantee that this will always work.

Literature

McLean IW, Nakave PK (1974) Periodate-lysin-paraformaldehyde fixative, a new fixative for immunoelectron microscopy. J Histochem Cytochem 22: 1077–1083

Newman GR, Jasani B, Williams ED (1982) The preservation of ultrastructure and antigenicity. J Microsc 127: RP5–RP6

Somogyi P, Takagi H (1982) A note on the use of picric acid paraformaldehyde-glutaraldehyde fixative for correlated light and electron microscopic immunocytochemistry. Neuroscience 7: 1779–1783

2.1.2 Dehydration

After the fixations and washing(s) have been completed, the specimens are embedded in a suitable plastic. Since such media are normally not miscible with water, this necessitates a dehydration step in which water is removed and replaced by a medium in which the plastic can be dissolved. Usually acetone or ethanol are employed for this purpose, the specimens being transferred stepwise through a concentration series, as given below:

	Duration (min)
Wash (in buffer or H_2O)	15
30% (v/v) ethanol or acetone	10
50% (v/v) ethanol or acetone	10
70% (v/v) ethanol or acetone	10
90% (v/v) ethanol or acetone	10
96% (v/v) ethanol or acetone	10
100% (v/v) ethanol or acetone	10
100% (v/v) ethanol or acetone	15

A useful generalisation for dehydration is that the more quickly the transfer into plastic has been executed the smaller is the chance that proteins or lipids will be leached out. Of the two dehydrating agents acetone tends to extract more lipid than alcohol. This is, however, not noticeable in the majority of ultrastructural investigations. If ethanol, however, has been used as a dehydrating agent the specimens have to be transferred to a second solvent, since the majority of embedding media do not dissolve in alcohol.

The dehydration steps as well as the gradual impregnation with plastic can be carried out at room temperature in glass vials or centrifuge tubes (both should be covered, either with a plastic cap or a glass marble), depending on whether one is dealing with tissue blocks or a cell suspension. Solutions are changed by pipetting (if necessary after centrifugation in a bench centrifuge – normally 3 min at full speed is sufficient to build a pellet). Care must be taken to avoid desiccation of the specimens during the dehydration; they must always be covered with liquid.

2.1.3 Embedding

A number of plastics are employed as embedding agents in EM. The most frequently encountered are the epoxy resins, although for certain purposes polyester and methacrylate mixtures may also be used. As a rule the transition from a monomeric, liquid state to a polymeric, solid state requires the presence of so-called curing as well as "modifying" agents. The former group includes "hardeners", which cross link the monomeric components, and "accelerators", which promote interactions between monomers. Modifying agents are also required to increase the tensile strength (i.e. to reduce brittleness) of the hardened plastic. They encompass substances which may become incorporated into the cross-linked product ("flexibilizers") and others which do not ("plasticizers").

Cross-linking reactions ("polymerisation") are slow at room temperature and require the addition of energy to reach completion in a reasonable period of time. This is given in the form of heat or by exposing to UV light. The latter has advantages for immunocytochemistry since it can be performed in the cold.

2.1.3.1 Embedding Media: General Usage and Precautions

The great majority of embedding media are toxic and can cause allergic reactions. When working with them it is necessary to wear plastic or rubber gloves and to carry out all manipulations in a fume cupboard. The various components for a particular embedding medium are normally stored in sealed bottles at 4° C. Before being opened, they must be allowed to reach room temperature in order to prevent contamination with condensed water. Embedding media are prepared either volumetrically or gravimetrically, each component being measured in an appropriate plastic vessel. The components are mixed with the help of a glass rod or stirrer bar and then allowed to stand for at least 10 min to allow air bubbles to escape. The final product should be clear and homogeneous.

The stability ("pot-life") of an embedding medium can vary considerably. As a rule, the epoxy resins are the most stable; they can be stored at 4° C for several days and at −20° C for several weeks or months. For storage purposes large amounts of embedding medium should be divided up into a number of small screwcap polyethylene bottles or small plastic syringes filled without leaving an air space. A fresh bottle should be thawed out for each new embeddment.

As with fixatives, used plastic embedding media should be collected in special vessels. If concentrated (i.e. without acetone), such waste plastic can be polymerized by heating before disposal. Glassware can be cleaned by rinsing with acetone, although in some cases it might be cheaper to throw the glass article away.

2.1.3.2 Conventional Embedding

2.1.3.2.1 Spurr's Medium. The most frequently used embedding medium at the moment is that introduced by Spurr (1969). It has a low viscosity (60 cps at 25° C) enabling rapid infiltration (see below) and can be polymerised in less than 24 h. It is made up of four components:

1. ERL 4206: Vinylcyclohexene dioxide (epoxy monomer).
2. DER 736: Diglycidyl ether of polypropylene glycol (flexibilizer).
3. NSA: Nonenyl succinic anhydride (hardener).
4. DMAE (S-1): Dimethyl aminoethanol (accelerator).

Table 2.2. Composition of Spurr's embedding medium

Component	Standard mixture	Hard mixture	Soft mixture
ERL 4206	10.0	10.0	10.0
DER 736	6.0	4.0	8.0
NSA	26.0	26.0	26.0
DMAE (S-1)	0.4	0.4	0.4
Total (in g)	42.6	40.4	44.4

Hardness is controlled by varying the relative amount of flexibilizer as given in Table 2.2.

2.1.3.2.2 Araldite. This medium has only three components:

1. Araldite CY212 (epoxy monomer).
2. DDSA: Dodecenyl succinic anhydride (hardener).
3. DMP30 (DY 064): 2, 4, 6 tridimethylamino methyl phenol (accelerator).

These are mixed together in the proportions given in Table 2.3. Araldite embedding media have the great disadvantage that they are very viscous (at least ten fold more than Spurr's medium) and the viscosity tends to increase with time (araldite becomes impossible to use after 2–3 days at room temperature).

2.1.3.2.3 Epon. Epon was widely used in the period from 1960 to 1975 and still enjoys use in many laboratories. In its most usual form an epon embedding mixture has four components:

1. Epon 812 (Epicote) (epoxy monomer).
2. DDSA: Dodecenyl succinic anhydride (hardener).
3. MNA: Methyl nadic anhydride (hardener).
4. DMP-30 (DY 064): 2, 4, 6 tridimethylamino methyl phenol (accelerator).

These are mixed together in the following proportions given in Table 2.4. The hardness of the blocks can be varied by changing the relative amounts of the two hardeners. The viscosity of the epon mixture depends on the composition but usually lies between 100 and 200 cps.

Table 2.3. Composition of Araldite embedding medium

Component	Amount (g)
Araldite	29.0
DDSA	24.0
DMP 30	0.5
Total	53.5

Table 2.4. Composition of Epon embedding medium

Component	Amount (g)
Epon 812	16
DDSA	8
NMA	8.7
DMP-30	0.4
Total	33.1

2.1.3.2.4 Infiltration and Polymerisation. The epoxy resins (Spurr's, Araldite, Epon) are soluble in acetone but not in alcohol. If, therefore, ethanol was employed as a dehydrating agent, a transitional solvent, e.g. propylene oxide (PO) must be used. In such cases the specimens are taken through two or three changes (each of 15 min duration) of 100% PO. Infiltration of the embedding medium occurs by gradually replacing the solvent with the liquid embedding medium. A typical schedule for the epoxy resins is as follows:

	Duration (h)
Acetone (100%) plastic 3:1 (v/v)	1
Acetone (100%) plastic 1:1 (v/v)	1
Acetone (100%) plastic 1:3 (v/v)	1
Acetone (100%) plastic	1
Acetone (100%) plastic	12
Acetone (100%) plastic	4

For the more viscous epoxy resins, e.g. Araldite, it may be necessary to include an extra step at the beginning, e.g. 6:1, solvent: plastic, and to extend the infiltration periods for each step as well. It may also be necessary to conduct the infiltration on a slowly rotating shaker.

The polymerisation conditions vary according to the type of embedding medium involved. Table 2.5 gives values for the major resins in use. After polymerising, it is good practice to allow the blocks to cool to room temperature for several hours before attempting to cut sections. The quality of an embeddment can often be judged by simple means: good blocks can just about withstand the pressure of a finger nail; good blocks tend to have a yellowish colour (if they have a brown colour this means that too much hardener was present and this can mean that the blocks are too brittle).

Table 2.5. Polymerisation conditions for epoxy-type resins

Plastic/resin	Temperature (°C)	Duration (h)
Spurr's	70	8-12
Epon	60	24-72
Araldite	a) 45	5
	b) 65	48

2.1.3.3 Water-Soluble Embedding Media

There is often the wish to avoid leaching and denaturing effects which may occur during conventional dehydration. In such cases one resorts to water-miscible embedding media which also have the extra advantage of reducing embedding-related cellular volume changes to a minimum. One of the most frequently used media of this type is Durcupan ACM (Fluka), an Araldite which replaces Epon 812 in an epon-like medium (Kushida 1964). The components and their proportions are given in Table 2.6. As with Epon, block hardness is controlled by the relative amounts of DDSA and MNA in the medium.

Table 2.6. Composition of Durcupan R-embedding medium

Component	Amount (ml)
Durcupan ACM	5.0
DDSA	1.7
MNA	1.0-1.2
DMP-30	0.2-0.4
Total	7.9-8.3

With Durcupan, dehydration and infiltration are carried out simultaneously at room temperature on a slowly rotating shaker. The following schedule is to be recommended:

	Duration
50% (v/v) Durcupan ACM in dist. H_2O	15-30 min
70% (v/v) Durcupan ACM in dist. H_2O	15-30 min
90% (v/v) Durcupan ACM in dist. H_2O	15-30 min
Durcupan ACM	30-60 min
Durcupan ACM	30-60 min
Durcupan ACM/Durcupan medium (3:1) (v/v)	1 h
Durcupan ACM/Durcupan medium (1:1) (v/v)	1 h
Durcupan ACM/Durcupan medium (1:3) (v/v)	1 h
Durcupan medium	1- 2 h
Durcupan medium	1- 2 h
Polymerisation at 50° C	50 h

A major problem with Durcupan embedded material is that it is difficult to section. This can be overcome by using Durcupan ACM as a dehydration agent, but then transferring the specimens to conventional embedding media (Spurr's, Araldite, Epon) before the polymerisation. A similar infiltration schedule can be employed, e.g. Durcupan ACM: epoxy medium (3:1, 1:1, 1:3; v/v). It is imperative that all traces of water-soluble Durcupan ACM in the final polymerisation mixture be removed prior to polymerisation.

2.1.3.4 Embedding for Immunocytochemistry

Conventional embedding media (Spurr's, Epon) have the tendency to give rise to unspecific background labelling when treated with ferritin- or gold-tagged antibody solution (see Sect. 2.4.2.2). A number of other media have been introduced over the last few years which appear to be much more suitable. These are presented in detail below.

2.1.3.4.1 Glycol-Methacrylate. This is a clear, colourless liquid of low viscosity. Since it is miscible with water, a conventional dehydration via acetone or water can be dispensed with. The specimens can thus be taken through increasing concentrations of aqueous Glycol-Methacrylate (2-hydroxyethyl methacrylate). The infiltration begins with an 80% aqueous solution of Glycol-Methacrylate for 30 min followed by a 97% solution of the same, also for 30 min. The specimens are then transferred to an embedding medium of the following composition:

Glycol-methacrylate (97%; v/v; solution in double distilled H_2O)	70 ml
Butyl-methacrylate + Benzyl Peroxide (2%; w/v)	30 ml
Divinyl Benzene (50%; v/v; in styrene)	10 ml

The embedding medium should be renewed twice over a 2 h period. The specimens are placed in gelatine (not polyethylene!) capsules and the capsules topped up with fresh embedding medium. Polymerisation is achieved by exposing to UV light (366 nm) for 2–3 days or, when antigenicity is not lost by high temperatures, by heating to 70° C for 12 h.

Sections of Glycol-methacrylate, as is the case with all methacrylates, are relatively unstable to the electron beam. Care must therefore be taken when working with sections of this type.

2.1.3.4.2 Lowicryl Resins. This type of embedding material is used quite often for immunocytochemical investigations. Not only does it allow good structural preservations, but the fact that the polymerisation with UV light can also take place at temperatures below 0° C ensures that antigenic properties are also well preserved.

Lowicryl resins are extensively cross-linked acrylate and methacrylate plastics. Two types are currently in use: K4M and HM 20. Both have low viscosities, even at low temperatures. Thus K4M can be used down to −35° C and HM 20 down to −70° C. Whereas HM 20 is nonpolar (hydrophobic), K4M is polar, which means that a certain amount of water can be tolerated during the infiltration process. It is even possible to polymerise K4M in the presence of 5% (w/w) H_2O.

The composition of the two types of Lowicryl embedding media is given in Table 2.7. Block hardness can be controlled by varying the relative amounts of cross linker (the more cross linker, the harder the block). The initiator C is necessary for polymerisations carried out below 0° C. Above 0° C it is replaced by the same amount of benzoin-ethyl ether. The mixtures should be mixed well but not extensively, since this can have a negative effect on the quality of the polymerisation.

Table 2.7. Composition of Lowicryl embedding media

K4M Component		Amount (g)[a]	HM 20 Component		Amount (g)[a]
Cross linker	A	2.7	Cross linker	D	2.98
Monomer	B	17.3	Monomer	E	17.02
Initiator	C	0.1	Initiator	C	0.10

[a] For UV polymerisation.

It is usual that, if the embedding process is to be carried out below 0° C, the dehydration should also be performed at low temperature. A suitable schedule, applicable for both types of media, is as follows:

Ethanol (vol %)	Temperature (° C)	Duration (min)
30	0	30
50	−20	60
70	−35 (−50)	60
95	−35 (−50 bis −70)[a]	60
100	−35 (−50 bis −70)[a]	60
100	−35 (−50 bis −70)[a]	60

[a] Values in brackets for HM 20 only.

For the infiltration the following schedule can be recommended:

Embedding medium: Ethanol (v/v)	Temperature (° C)	Time (h)
1:1	−35	1
2:1	−35	1
100% medium	−35	1
100% medium	−35	4–16

lower temperatures are possible when working with HM 20.

The low temperatures can be achieved with the appropriate cryogenic mixtures, e.g. for −20° C: ice-NaCl mixture of 3:1 (w/w); for −30° C to −70° C: o- and m-xylol in dry ice in varying proportions. More reproducible results are obtained with household freezers. A machine for this purpose is commercially available from Balzers.
For the infiltration the following schedule can be recommended:
The specimens are then transferred to prefilled, precooled capsules with a precooled Pasteur pipette. Wait until the temperature has equilibrated (10–15 min) before starting the polymerisation.

Polymerisation can be carried out in polyethylene or gelatine capsules. Important is that specimens are exposed to UV-light (360 nm) from all sides. 15 W flourescent tubes, as used in thin layer chromatography, will suffice for this purpose. When self-made polymerisation chambers are used, provision must be made for varying the lamp-capsule distance as well as the temperature. In addition, the air in the chamber must be made to circulate. The polymerisation takes 24 h at a temperature of $-30°$ to $-40°$ C. Before being sectioned, the blocks should be stored for 2–3 days at room temperature.

A polymerisation at 60° C (without UV light) is also possible. For this, gelatine capsules should be used. It is also necessary that the extra heat which is developed in this case, as a result of the polymerisation reactions, be removed. This can easily be done by inserting the capsules into holes in an aluminium block.

Although embedding with Lowicryl resins is usually carried out at low temperatures, it has been shown (Altman et al. 1984) that antigenicity in some cases is still retained when specimens are fixed, dehydrated and embedded in Lowicryl at room temperature and polymerised with UV light at 4° C.

2.1.3.4.3 London Resins (LR). LR-white and LR-gold are acrylates manufactured by the London Resin Company. Both have relatively low viscosities and penetrate cells/tissues easily. They are polar and this hydrophilic property makes them very suitable for immunocytochemical purposes. Because of their capacity to take up water it is not necessary to completely dehydrate specimens with organic solvents before infiltrating with the plastic. The decision to use LR-white or LR-gold depends on the properties of the antigen in question. LR-white can be polymerized at 50° C in 24 h or at $+4°$ C in the presence of a light-sensitive initiator (Benzil) by UV light (100-W halogen lamp; 20 cm distant). Many antigens lose their reactivity/specificity at 50° C; on the other hand, the initiator Benzil can alter antigenicity. Low temperature ($-4°$ C) embedding with LR-white for immunocytochemistry is not recommended, since the presence of the accelerator stimulates exothermic reactions, leading to very high local temperatures which cannot always be adequately corrected for. Embedding schedules for the two plastics are as follows:

LR-White

1. Dehydrate cells/tissue blocks in 70% (v/v) acetone/ethanol for 15 min.
2. Transfer to LR-white; leave for 1 h at room temperature; repeat.
3. Transfer to gelatine capsule; fill completely with LR-white (leave no air space - O_2 inhibits the polymerisation)
4. Polymerise at 50° C for 24–36 h.

LR-Gold

1. Dehydrate cells/tissue blocks in 70% (v/v) acetone or alcohol for 15 min.
2. Transfer to LR-gold at $-25°$ C; leave for 1 h.
3. Transfer to fresh LR-gold containing 0.5% (w/v) Benzil; leave overnight (in the dark) at $-25°$ C.

4. Transfer to gelatine capsules; fill completely with LR-gold + 0.59% (w/v) Benzil at −25° C.
5. Polymerise with UV light (100-W Osram Xenophot halogen lamp, 20 cm distant) for 24–36 h at −25° C.

Self-made or commercial (Balzers) low temperature embedding apparatuses can be employed, as is the case with Lowicryl resins.

2.1.3.4.4 Embedding in Gelatine for Cryo-Ultramicrotomy. In contrast to cells in tissues, the cells in cell suspensions must somehow be kept in place during the various steps in cryo-ultramicrotomy (see Sect. 2.2.5). This is best accomplished by embedding in gelatine. Gelatine embedding is also sometimes used for tissues, since trimming is made easier and the surface tension problems which arise during the thawing of cryosections (on the sucrose drop) are considerably reduced when a gelatine matrix is present.

The following embedding procedure is recommended for bacteria and algal cells: prefixed cells are resuspended in 10% (w/v) gelatine (made up in a compatible buffer, e.g. 0.1 M sodium phosphate) at 37° C. The cells are centrifuged out of the gelatine solution and poured onto a glass slide to produce a 1-mm-thick layer which is solidified by cooling to 4° C. A second primary fixation with buffered aldehydes is then carried out, followed by several wash steps in the same buffer. Finally, small blocks are cut out of the solidified gelatine-cell mass. These blocks can then be frozen and processed for cryosectioning.

2.1.3.5 Embedding Moulds and Specimen Orientation

As already mentioned, embedding can take place in gelatine or polyethylene (e.g. BEEM or TAAB) capsules. Such capsules constitute a "vertical" embedding. When a specific orientation is required it is often more convenient to use a flat embeddment. The embedding mould in such cases can range from small volume types in silicone rubber to large volume types, e.g. polyethylene weighing trays (boats) or self-made moulds of aluminium foil. The orientation of the specimens in flat forms is best carried out before the polymerisation is started. If this is not possible, most of the embedded objects can be cut out of the polymerised blocks with a fretsaw and stuck (with a commercial, quick-drying, plastic glue) onto a preformed support block of the same embedding material.

The necessary materials for the embeddment (capsules, pipettes etc.) should be completely dry before bringing them into contact with embedding medium. In order to know which specimen is which after polymerisation, it is necessary to embed a small piece of paper bearing some sort of code number(s) with each specimen. The writing can either be in pencil or done with a typewriter but not with a ball-point pen, the writing of which tends to dissolve during the polymerisation process.

2.1.3.6 Embedding of Monolayer Cell Cultures

The method used for embedding monolayer cell cultures depends to a great extent on the way in which the cells are grown. Thus cells grown on plastic dishes are much easier to work with as compared to those grown in plastic bottles. Most important is that the plastic surface used to support the cells does not interact in any way with either the dehydrating solvent or the embedding medium. In this respect one of the best types of plastic growth dishes is one made of Thermanox (Flow Laboratories Ltd.) which is not attacked by acetone, alcohol, styrene, propylene oxide or by epoxy resins. In addition, the majority of epithelial and mesenchymal cell lines grow excellently on Thermanox. Thermanox can also be embedded and trimmed just like normal tissue; however, in order to avoid mistakes a recognition marker should be made on the noncellular side of the Thermanox.

Interestingly, and probably because of the short diffusion pathways involved, it is not absolutely necessary to proceed from the dehydrating solvent (acetone or propylene oxide) to the embedding medium via a stepwise dilution series as described above (Sect. 2.1.3.2.4). One can simply add the embedding medium direct to the monolayer after the last dehydration step. If this is done the cells should be left for at least 12 h (or overnight) in the embedding medium before polymerisation.

Since the embedding medium does not interact with the support surface (Thermanox, plastic Petri dishes, plastic bottles) there is a clear-cut border between the two types of plastic after polymerisation. The embedded cells, together with the plastic support, can be sawed out and trimmed as for normal flat embedding (see Sect. 2.1.3.5). During this process the plastic support usually breaks off, leaving a smooth, flat, surface on the hardened embedding material. The cells lie adjacent to this surface and can be sectioned without problem.

Literature

Altman LG, Schneider BG, Papermaster DS (1984) Rapid embedding of tissues in Lowicryl K4M for immunoelectron microscopy. J Histochem Cytochem 32: 1217-1223

Armbruster BL, Carlemalm E, Chiovetti R, Garavito LRM, Hobot JA, Kellenberger E, Villinger W (1982) Specimen preparation for electron microscopy using low temperature embedding resins. J Microsc 126: 77-85

Carlemalm E, Garavito RM, Villinger W (1982) Resin development for electron microscopy and an analysis of embedding at low temperature. J Microsc 126: 123-143

Flickinger CJ (1966) Methods for handling small numbers of cells for electron microscopy. In: Prescott DM (ed) Methods in cell physiology, vol II. Academic Press, London New York, pp 311-321

Glauert AM (1975) Fixation, dehydration and embedding of biological specimens. In: Glauert AM (ed) Practical methods in electron microscopy, vol III, part 1. Elsevier/North Holland, Amsterdam, pp 1-207

Horvat R, Hovovka A, Dekan G, Poczewski H, Kerjaschki D (1986) Endothelial cell membranes contain podocalyxin - the major sialoprotein of visceral glomerular epithelial cells. J Cell Biol 102: 484-491

Kruse PF, Patterson MKL (1973) Tissue culture: methods and application. Academic Press, London New York, pp 1-868
Kushida H (1964) Improved methods for embedding with Durcupan. J Electron Microsc 13: 139-144
Plattner H (1981) Die Entwässerung und Einbettung biologischer Objekte für die Elektronenmikroskopie, Kap 2.3.1. In: Schimmel G, Vogell W (eds) Wiss Verlagsges, Stuttgart, pp 1-51
Roth J, Bendayan M, Carlemalm E, Villinger W, Garavito RM (1981) Enhancement of structural preservation and immunocytochemical staining in low temperature embedded pancreatic tissue. J Histchem Cytochem 29: 663-671
Spurr AR (1969) A low-viscosity epoxy resin embedding medium for electron microscopy. J Ultrastruct Res 26: 31-43
Yokota S, Oda T (1986) Immunocytochemical demonstration of serine: pyruvate aminotransferase in peroxisomes and mitochondria of rat kidney. Histochemistry 83: 81-85

2.2 Ultramicrotomy

2.2.1 Trimming of Blocks

2.2.1.1 General

Blocks can be trimmed to produce a flat pyramid form, either manually or mechanically. In the former case this can be done with a clean razor blade, theoretically with five cuts (see Fig. 2.9). The same can be achieved, more precisely, with a histological microtome. (LKB Pyramitome using a glass knife - see Sect. 2.2.2 - in which 5-μm-thick sections are made at an angle to the main axis of the block) or a rotating milling cutter (Reichert TM 60 specimen trimmer fitted with either a steel cutter, which gives rise to rough surfaces, or a diamond cutter, which leaves a clear surface). Blocks can normally be trimmed more quickly with a milling cutter than with the Pyramitome; both, however, take longer than manual trimming in the hands of experienced workers.

The face of the pyramid should be flat and free from plastic debris. The form is most important; serial sections are produced most easily when the trimmed surface is in the form of a trapezium. A square form is also usable, but is not so reliable for a long series of serial sections.

Literature

Gugliemotti V (1976) Device for manual trimming of tissue blocks for ultramicrotomy. Stain Technol 51: 135-138

2.2.1.2 Controlled Trimming: Production and Staining of Semi-Thin Sections

It is often necessary to find and trim precisely around a particular part of an embedded object. In order to do this, semi-thin sections (0.5-2 μm) have to be made and examined either after staining or directly in a phase contrast micro-

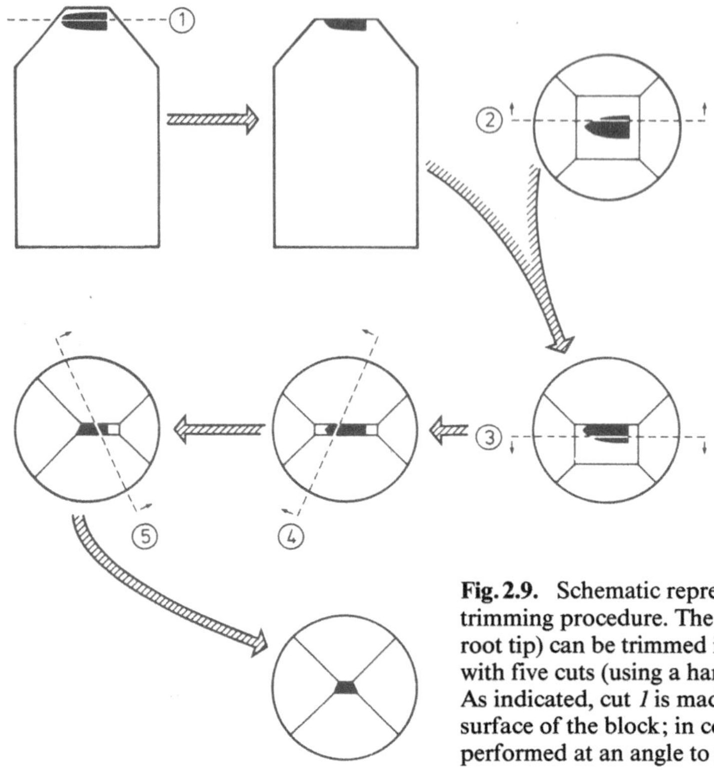

Fig. 2.9. Schematic representation of a block trimming procedure. The object (in this case a root tip) can be trimmed into a trapezium form with five cuts (using a hand-held razor blade). As indicated, cut *1* is made parallel to the flat surface of the block; in contrast, cuts *2–5* are performed at an angle to this surface

scope. The sections are transferred with a hair brush to a drop of distilled water on a glass slide. Through warming to 60–80° C (heating plate) they become stretched and dry down onto the slide surface. Higher temperatures may lead to folds and air bubbles in the sections. The sections are then stained with either toluidine blue or methylene blue azure.

Toluidine Blue Staining

a) *Solution:* 0.5–3% (w/v) Toluidine blue in 0.1 M K-phosphate buffer. The solution should be heated to 90° C for 15 min and then filtered.
b) *Staining procedure:* The sections on the microscope slide are covered with the stain and then warmed to 50° C for 5 min. During this period it is important that an evaporation of water from the solution be prevented in order to avoid an unspecific precipitation of stain. The sections are then washed with distilled water to remove unreacted stain. If the specimens are too deeply stained, excess stain can be removed by extracting with 70% (v/v) ethanol. The extent, depth and intensity of the staining can be varied by adjusting the stain concentration or extending the length of the staining period.

Azure-II - Methylene Blue Staining

a) *Solutions:* (1) 1% (w/v) periodic acid in distilled water (100 ml). (2) 1% (w/v) sodium tetraborate plus 1% (w/v) methylene blue in 50 ml distilled water. (3) 1% (w/v) azure-II in 50 ml distilled water. Solutions 2 and 3 are mixed 1:1 (v/v) and 50 g sucrose is added.
b) *Staining procedure:* Add several drops of the periodic acid solution to the sections and allow to stand for 5 min. Wash off and dry the borders of the glass slide. Add several drops of the stain to the sections and warm to 60-80° C (heating plate) for 5-15 min. Quickly wash the stain off the sections and dry on the heating plate. Cover the sections with Caedax or paraffin oil. Examine immediately because the staining tends to decrease in intensity with time.

Literature

Lewis PR, Knight DP (1977) Staining methods for sectioned material. In: Glauert AM (ed) Practical methods in electron microscopy, vol V, part 1. Elsevier/North Holland, Amsterdam, pp 1-311

Richarson KC, Jarret L, Finke RH (1960) Embedding in epoxy resins for ultrathin sectioning in electron microscopy. Stain Technol 35: 313-323

2.2.2 Preparing Glass Knives

The introduction of the glass knife by Latta and Hartmann (1950) represented a significant step forward in thin sectioning technique. Essentially, glass knives are produced by breaking off quadratic pieces from a strip of glass, scoring them diagonally and then breaking into two triangular knives (see Fig. 2.10). Two angles are to be distinguished in the final breakage: the scored angle (δ) and the knife angle (α). The scored break should be arranged so that the difference between the two angles lies between 5° and 15°.

Originally, glass knives were made with the help of a glass scorer and a pair of pliers modified with strips of rubber to provide three points of contact. Today, glass knives are normally made with a piece of apparatus specially constructed for this purpose. The most frequently encountered type is the LKB knife-maker 7800 B, the use of which is described below.

2.2.2.1 Preparation of the Glass Strips

Glass strips (0.6 cm thick) are usually available in widths of 25 or 38 mm (length 40 cm). The strips should be cleaned in the following manner:

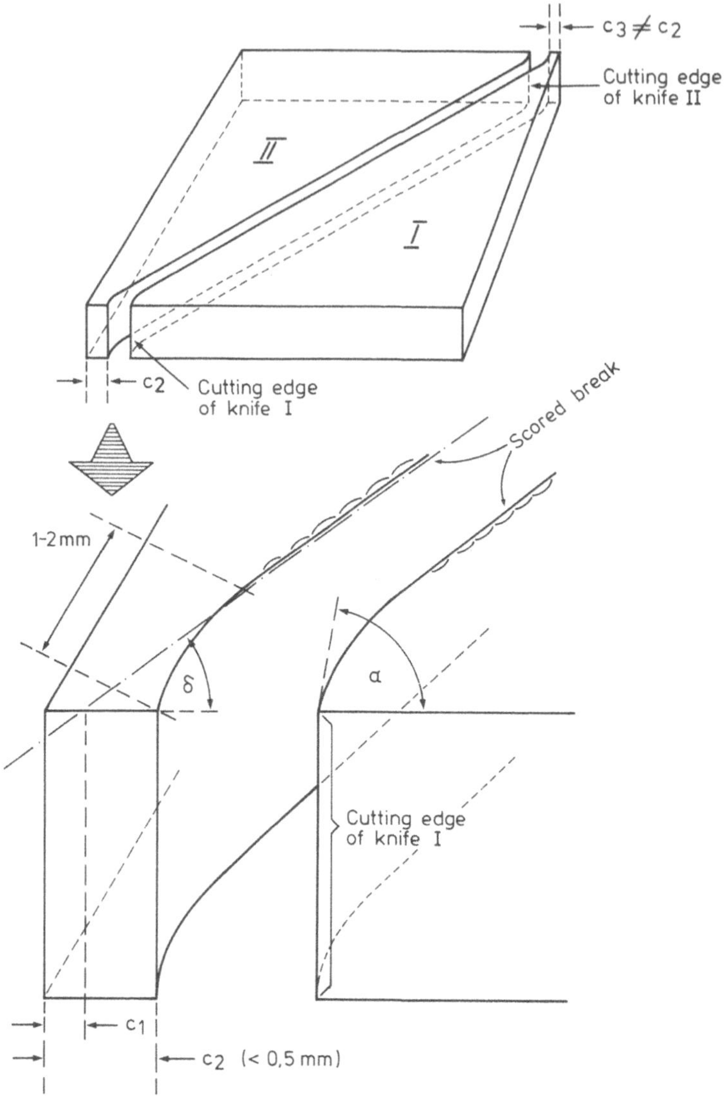

Fig. 2.10. Preparation of a glass knife. *Above:* breaking a glass rectangle into triangles. *Below:* geometrical parameters which arise during the production of a cutting edge

- Wash with soapy water.
- Rinse in running water.
- Rinse with double-distilled water.
- Clean with cotton wool pad soaked in ethanol.
- Dry with a clean handtowel.

After this the sides of the glass strip should not be touched with bare hands.

2.2.2.2 *Breaking Glass Squares*

Preliminary Adjustments on the Knife Maker

- Set the clamping lever in the rear position.
- Bring the angle setting plate into position.
- Turn the breaking knots fully anticlockwise.
- Push the scoring tool into the clamping head.

Breaking Procedure

- Lay the glass strip on the base plate with one edge resting on the angle-setting plate. Push to the left until stopped by the arresting studs.
- Slowly lower the clamping head by setting the clamping lever in the forward position. Hold the glass strip with the right hand until the clamping head touches the glass then take hand away and press clamping lever fully downwards.
- Introduce "collecting fork" underneath the glass strip from the left.
- Pull the scoring tool out of the clamping head so that the scoring wheel moves smoothly but distinctly across the glass.
- Turn the breaking knob clockwise until the glass strip breaks, then turn the knob back anticlockwise.
- Hold the glass strip with the right hand and raise the clamping head by pushing the clamping lever to the rear.
- Remove glass strip and lift the glass square with the collecting fork.
- Place the glass squares onto a clean handtowel.

2.2.2.3 *Making Knives*

Preliminary Adjustment to the Knife Maker

- Pull back rubber clamping pads.
- Set front sliding glass holder to "9".
- Set rear sliding glass holder to "6".
- Set front lateral adjustment knob to "5".
- Set rear lateral adjustment knob to "1".
- Insert a glass square on the breaking plate with the help of the collecting fork.
- Bring rear sliding holder forwards with the release knob until it touches a corner of the glass square, then retract and clamp.
- Move rear sliding holder two divisions on the scale forwards and clamp.
- Select "25" and ◇ on the scoring tool.
- Lower clamping head and press downwards.
- Pull the scoring tool out of the clamping head.
- Raise clamping head.
- Remove scored glass square and compare the scored break with the ideal (see Fig. 2.10).

Breaking Procedure

- Bring rubber damping pads into position "O".
- Insert glass square with collecting fork underneath.
- Bring rear sliding holder forwards; bring front sliding holder into position until the glass square is held tightly (as above).
- Bring down damping head and score.

- Turn the breaking knob clockwise until the glass strip breaks and then turn back anticlockwise. (The quality of the knife is determined in part by the breaking speed - slow breakages give the best knives.)
- Retract damping pads; raise clamping head; push scoring tool back into clamping head; retract and clamp rear sliding holder.
- Carefully remove glass knives with the fork (the broken edges should not be allowed to touch one another).

2.2.2.4 *Judging the Quality of a Glass Knife*

A good guide to determining the quality of a glass knife is the presence of a clear stress (Wallner) line leading from the left-hand corner of the cutting edge (Fig. 2.11). Glass knives are best inspected by clamping them in the knife holder of an ultramicrotome and adjusting the illumination so that the cutting edge is seen as a narrow white band against a darker background. Good knives should not show any irregularities when viewed at magnifications of between 20 × and 50 ×. The extent of the cutting edge which can be used for sectioning (Z in Fig. 2.11) depends on the angle of the knife and the method of breaking, but it is usually restricted to about ⅓ of the total.

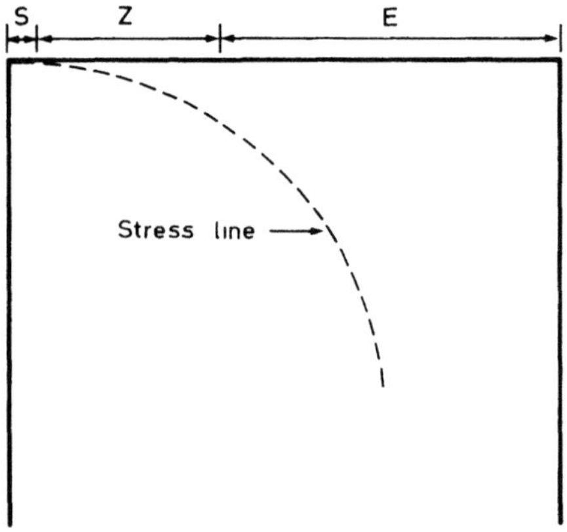

Fig. 2.11. The cutting regions of a glass knife. *S* the corner which should not be used for sectioning; *Z* the region of the cutting edge used for obtaining sections; *E* this region, depending on its quality, can also occasionally be used for sectioning

2.2.2.5 *Attaching Troughs*

Troughs must be attached to glass knives in order to hold water, which is necessary for collecting the sections as they are being cut. There are two types of trough: preformed ones made of metal or plastic, or ones made from reflecting sticky tape (e.g. gold-silver Scotch tape). Troughs of the latter type are simple to construct:

- Cut a piece of Scotch tape (3-4 cm).
- Place the glass knife on the sticky side of the Scotch tape so that the cutting edge of the knife lies parallel to the edge of the tape.

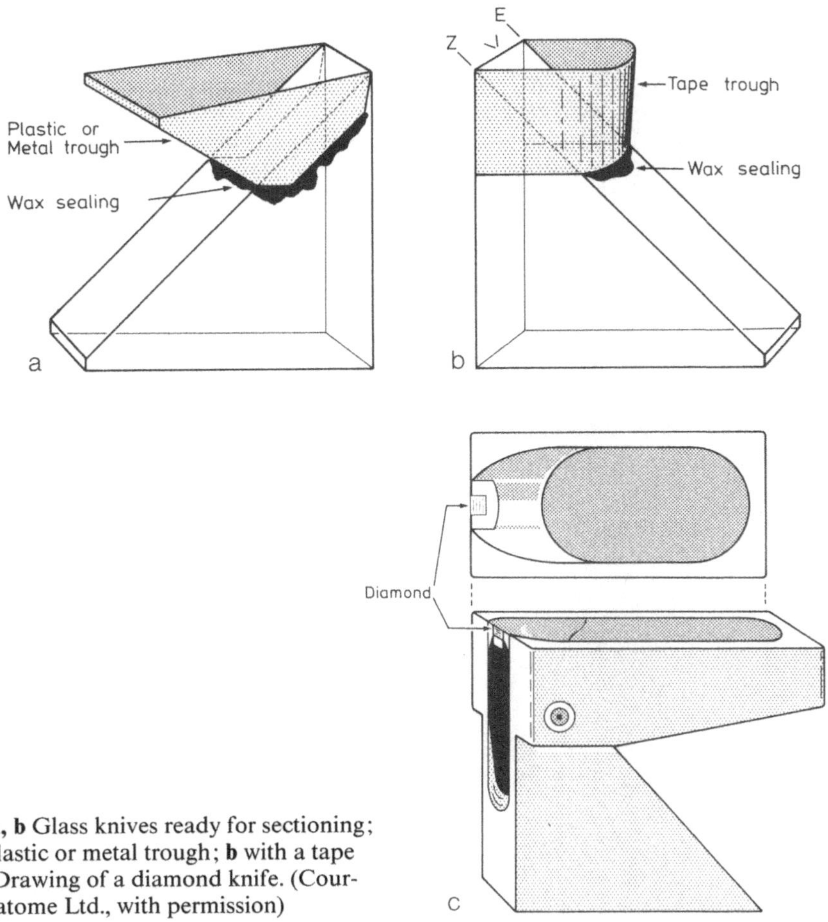

Fig. 2.12. a, b Glass knives ready for sectioning; **a** with a plastic or metal trough; **b** with a tape trough. **c** Drawing of a diamond knife. (Courtesy of Diatome Ltd., with permission)

- Pick up the glass knife together with the tape. Bend the tape around the knife as shown in Fig. 2.12b and press onto the other flat surface of the knife.
- Carefully cut off the excess tape with a razor blade (cutting from the bottom in the direction of the knife edge).
- Seal the trough with molten dental wax from outside (this also has to be done for preformed troughs).

2.2.2.6 *Storing Glass Knives*

Glass knives should be used within days of their having been made. During this period they must be stored in a dust-free container with their cutting edge upwards, i.e. not on their sides. If dust particles are present in the trough they are removed by squirting double-distilled water several times into and out of the trough. Glass knives are not often used more than once. If they are, it is important that a new portion of the cutting edge is used each time.

Literature

Latta H, Hartmann JF (1950) Use of glass edge in thin sectioning for electron microscopy. Proc Soc Exp Biol Med 74: 436-439
Sheldon H (1957) A method for evaluating glass knives. J Biophys Biochem Cytol 3: 621-624
Weiner S (1959) A new method of glass knife preparation for thin section microtomy. J Biophys Biochem Cytol 5: 175-177

2.2.3 Diamond Knives and Their Care

Diamond knives were introduced by Fernandez Moran (1953) and are particularly useful when sectioning hard objects (e.g. bony or woody tissue). Moreover, because of their extreme sharpness and long life, they are to be preferred to glass knives. A disadvantage, however, is their price and the cost of resharpening.

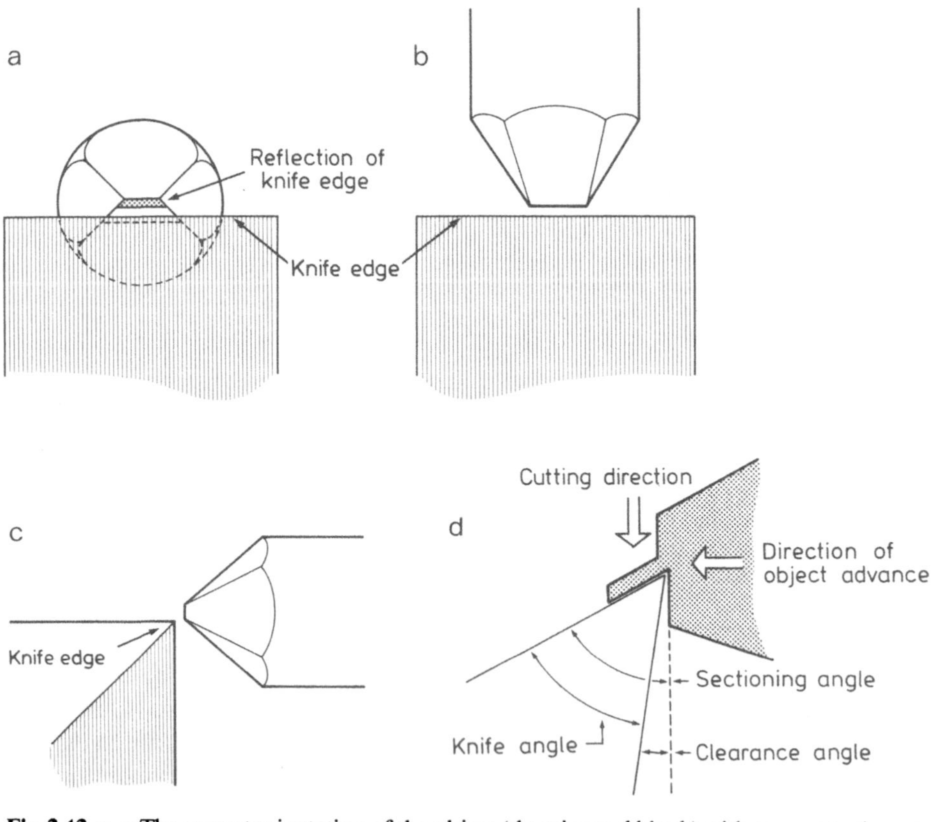

Fig. 2.13. a-c The correct orientation of the object (the trimmed block) with respect to the cutting edge of the knife as seen from three different orientations. **d** Geometrical parameters characterising sectioning

The diamond itself is sintered into a metal shaft, which in turn is set into a metal holder which includes a trough (Fig. 2.12 c). Diamond knives are supplied by many companies (e.g. Du Pont, Diatome, Balzers) and the manufacturers' instructions as to the usage, in particular the clearance angle (see Fig. 2.13 d), should always be heeded. Although significant differences in the usage of diamond as against glass knives do not exist, no one should attempt to learn sectioning starting with a diamond knife. Care must also be taken when stretching sections (see Sect. 2.2.4.3), since the chloroform-soaked cotton wool pad can cause the sinter around the diamond to dissolve away.

Occasionally it becomes necessary to clean the cutting edge of a diamond knife, since small particles of embedding plastic begin to adhere with time. This is done with a stick of elder pith shaped at one end into a chisel form. The tip of this stick is soaked in ethanol and pushed carefully onto the knife edge.

Literature

Fernandez Moran H (1953) A diamond knife for ultrathin sectioning. Exp Cell Res 5: 255–256

2.2.4 Conventional Sectioning

2.2.4.1 Trough Liquids

Trough liquids must fulfill three requirements:

a) they should not react with the sections;
b) they should possess a high surface tension in order that the sections may float easily and that the liquid does not run over the knife edge;
c) they should not be volatile.
Water is thus an optimal trough liquid. During sectioning it is important that the water level be controlled and, if necessary, be adjusted. On the one hand, a convex meniscus can lead to wetting of the block surface as well as the back of the knife, both of which prevent the production of ribbons of sections. On the other hand, a concave meniscus can result in the sections piling up and becoming folded. One tries therefore to maintain a flat meniscus. To this end the majority of modern ultramicrotomes are fitted with a device to change the level of the water in the trough.

2.2.4.2 Using an Ultramicrotome

There are two different types of ultramicrotome: those in which the block advances onto the knife edge through the heat-induced expansion of the object holder (thermal advance), and those in which the block advance is achieved mechanically (e.g. the Du Pont MT 5000 and the Reichert Ultracut). Although each ultramicrotome should be operated according to the manufacturers' instruction, a number of steps are common to all:

- Switch microtome on.
- Switch light on and adjust accordingly.
- Insert knife, clamp into position at the correct clearance angle (Fig. 2.13 d).
- Insert trimmed block with the cutting surface parallel to the cutting edge of the knife (Fig. 2.13 a-c).
- Fill the trough with water.
- Start motor, at first with a small block advance. Check that the surface of the block passes the cutting edge of the knife during the second half of the downwards cutting movement.
- Adjust water meniscus in trough.
- Raise cutting speed, increase block advance until the desired section thickness is obtained.
- Stop the motor in order to pick up the sections; the block should then lie below the level of the cutting edge of the knife.

2.2.4.3 *Section Thickness*

Sections are usually produced which have a smaller surface area than the block face from which they have been taken; they are thus compressed. This artefact can be reversed by exposing the sections to the vapour of an organic solvent for a few seconds. This "stretching" is best achieved by holding a pad of chloroform-soaked cotton wool about 1 cm above the ribbon of sections.

Section thickness can be estimated on the basis of interference colours. The thinnest sections have a grey to silver grey colour (see Table 2.8).

Table 2.8. Determination of section thickness on the basis of their interference colour

Interference colour	Section thickness (nm)
Grey	Under 60
Silver	60- 90
Gold	90-150
Purple	150-190
Blue	190-240

2.2.4.4 *Picking Up Sections*

When sectioning has been performed properly, ribbons of sections are produced. If only individual sections are obtained then sectioning has not been done correctly (see Table 2.9). Serial sectioning is therefore not a special cutting technique but merely the careful collection of ribbons of sections without loosing or damaging an individual section.

The steps involved in picking up sections are presented diagrammatically in

Table 2.9. Troubleshooting in ultramicrotomy

Problem	Cause	Solution
1. The surface of the specimen has become wet	a) Too much water in trough	a) Less water in trough b) Dry specimen surface with filter paper; if necessary change block
2. Ribbons of sections will not form	a) Too much water in trough b) Upper and lower edges of the pyramid are not parallel c) Cutting speed too low	a) Reduce water level in trough b) Retrim block c) Increase cutting speed
3. Differences in thickness within individual section; sections in a ribbon with different thicknesses	a) Block too soft b) Knife and/or block not properly clamped c) Knife is blunt d) Incorrect knife position e) Cutting speed too high f) Cutting surface of block is too large	a) Repolymerise b) Clamp knife/block c) Move to another region of the cutting edge or change knives d) Check knife parameters (clearance angle etc.) e) Decrease cutting speed f) Retrim block
4. Sections fall behind the cutting edge of the knife	a) Block too soft b) Cutting speed too high c) Incorrect knife position	a) Repolymerise b) Increase cutting speed c) Check knife parameters
5. Sections have scratches in them	a) Cutting edge of knife defect b) Inhomogeneity in the block	a) Move to another region of the cutting edge or change knives b) Retrim or change block
6. Sections have holes in them	a) Air bubbles in the block b) Inhomogeneity in the block	a) Retrim b) Retrim or change block
7. Sections have chatter (see Fig. 2.15)	a) The block hits the knife and causes vibrations b) Inhomogeneous embedding	a) Replace block with a softer one; reduce cutting speed; trim a smaller pyramid; reduce clearance angle b) Re-embed with a different plastic

Fig. 2.14a–c. Ribbons which are too long for a grid are separated with the help of a mounted eyelash. One should note that the sections are not touched directly in this process, instead gentle currents in the trough liquid are created, which cause ribbons of sections to break at a particular point. After positioning the short ribbons of sections on the water surface, the sections can be picked up.

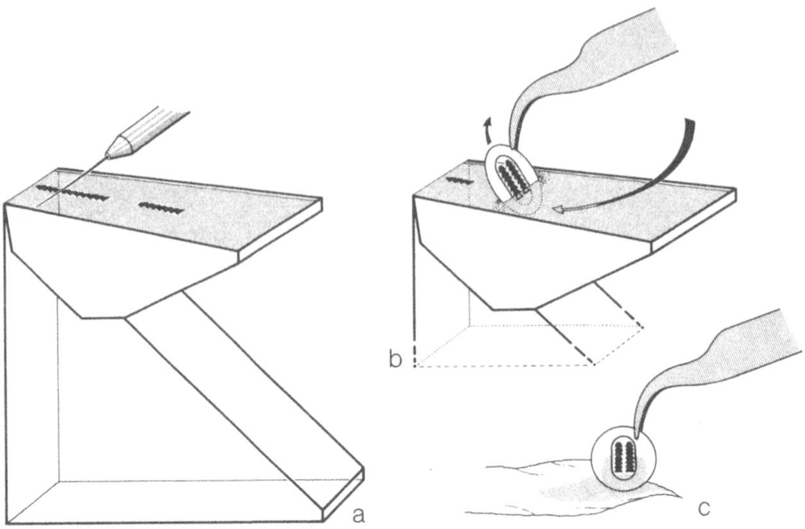

Fig. 2.14 a–c. Picking up sections. **a** separating a ribbon of sections with the help of a mounted eyelash; **b** picking up sections from below (for simplicity a single slot grid is shown here). **c** drying a grid with a piece of filter paper

Sections are better picked up from below, since their position on the grid can be controlled more easily. They can, however, be picked up from above by slowly lowering the grid onto the surface of the water, although with this method the sections can occasionally develop wrinkles. Excess water is drained off the grid by holding the grid vertically on a piece of filter paper (Fig. 2.14 c).

2.2.4.5 Sectioning Problems

There are a number of factors which can affect the quality of the sections obtained:

1. the quality of the knife edge,
2. the angle of section (clearance angle plus knife angle),
3. the cutting speed,
4. the cutting pressure,
5. the advance,
6. the type and quality of the embedding medium,
7. the form and size of block surface.

Often problems in sectioning are the result of several factors, rather than a single one. Table 2.9 summarizes the major problems which can occur when sectioning, gives their causes and suggests suitable courses of action to alleviate the problem.

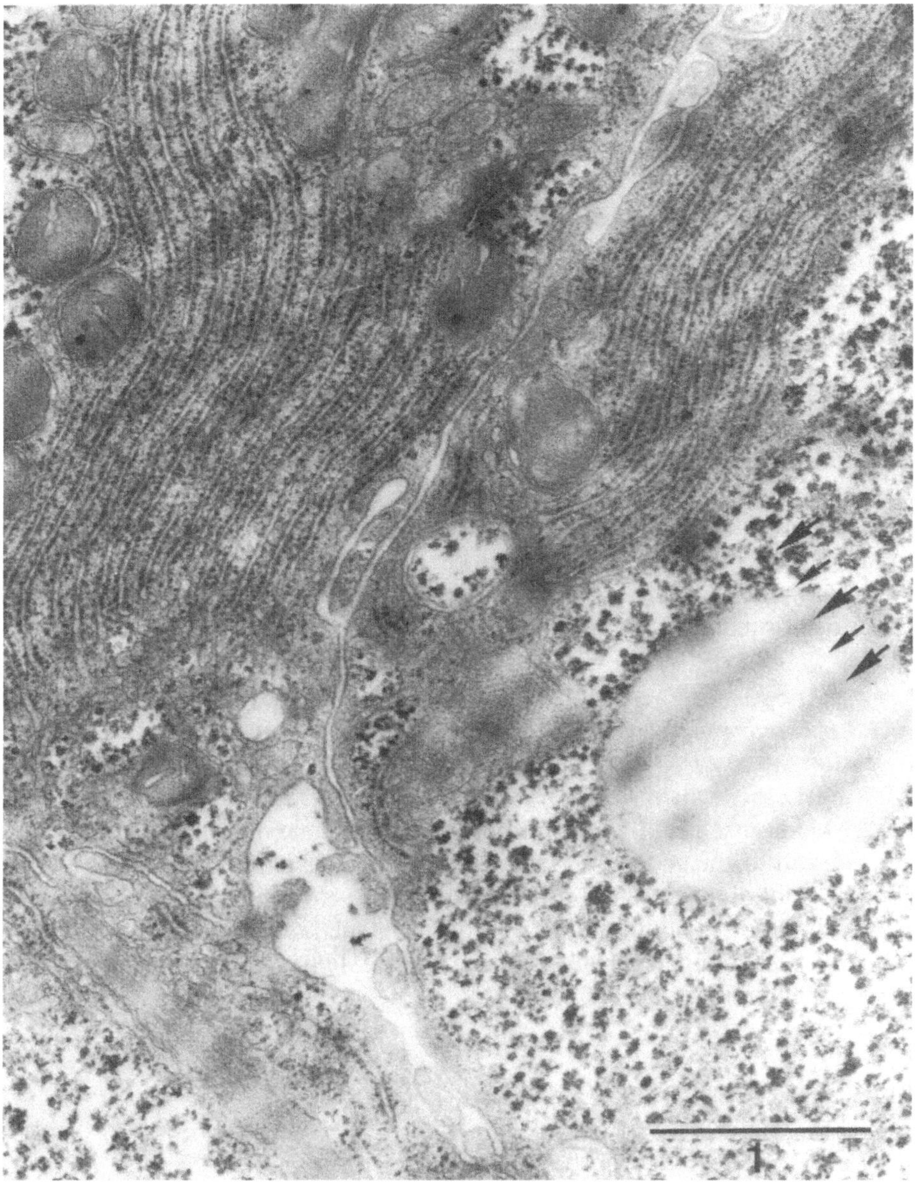

Fig. 2.15. "Chatter" in thin (grey interference colour) sections. This is an artefact (indicated by alternating *large and small arrows*) which is not usually noticed in the binocular microscope during sectioning. Object: rat hepatocyte. Magnification bar in μm

Literature

Hayat MA (1970) Principles and techniques of electron microscopy, vol I. Van Nostrand-Reinhold, New York
Reid N (1975) Ultramicrotomy. In: Glauert AM (ed) Practical methods in electron microscopy, vol III, part 2. Elsevier/North Holland, Amsterdam, pp 213-353
Sitte H, Neumann K (1983) Ultramikrotome and apparative Hilfsmittel für die Ultramikrotomie. In: Schimmel G, Vogell W (eds) Methodensammlung der Elektronenmikroskopie. Chap. 1.1.2. Wiss Verlagsges, Stuttgart, pp 1-248
Westfall JA (1961) Obtaining flat serial sections for electron microscopy. Stain Techol 36: 36-38

2.2.5 Cryo-ultramicrotomy

Cryo-ultramicrotomy differs from conventional ultramicrotomy in that thin sections are produced from cells embedded in an ice eutectic rather than in plastic. Since the specimens are not exposed to organic solvents or resins they retain much more of their natural character, e.g. antigenic properties than would otherwise be the case. Fixation and gelatine embedding, which usually precede this method, are dealt with elsewhere (see Sects. 2.1.1.7 and 2.1.3.3.4).

2.2.5.1 Freezing the Sample

A 1-mm^3 block (maximal size) is allowed to stand for 30 min in a solution of 2-2.3 M sucrose in 0.1 M sodium phosphate buffer pH 7.0. The block is then transferred to the specimen holder of the cryo-ultramicrotome and frozen by plunging into liquid freon or melting nitrogen (see Sect. 2.6.2.4); it can then be stored in liquid N_2 until required. Whilst this method has been successful with animal and bacterial cells, plasmolysis and ice damage prevent its being adopted for the majority of plant cells.

2.2.5.2 Sectioning the Frozen Sample

Glass knives are used for sectioning, as already discribed (see Sect. 2.2.2), but they should be of the best quality possible (see Griffiths et al. 1984 for a detailed procedure on how to obtain the "optimal" glass knife for cryosectioning). The glass knives are used without a trough and are kept dry. The cooled specimen holder carrying the frozen specimen block is transferred from liquid nitrogen to the cryochamber of the ultramicrotome which should have a temperature of between −90° C and −110° C. Advancing the block towards the knife edge is carried out as for conventional thin sectioning (see Sect. 2.2.4). Semi-thin (0.5-1 μm) sections are cut first followed by ultrathin (50-100 nm) sections.

2.2.5.3 *Picking up Frozen Sections*

An eyelash attached to a holder and a wire (platinum) loop (diameter 2 mm) are needed for the manipulation of the cryosections inside the cryochamber. The sections (they are often curled) can be straightened and removed from the dry glass knife with the eyelash. They are quickly touched, from above, with the surface of a drop of 2.3 M sucrose in phosphate buffer (see Sect. 2.2.5.1), which sits in the platinum loop. In this way the sections are transferred to a carbon-coated formvar grid with the sections becoming attached to the support film. The grid with the sections facing downwards is then placed on an ice-cooled plate covered with a thin layer of gelatine (2% w/v in phosphate buffer). The sucrose thus diffuses into the gelatine and is removed from the sections. The plate covered by gelatine is prepared by pouring liquid gelatine (at room temperature) into a small Petri dish kept a high humidity (moisture chamber) and then allowing the gelatine to solidifying by placing the Petri dish on ice. For immunolabelling of the sections see Section 2.4.2.2; for uranyl staining see Section 2.2.6.4.

Literature

Griffiths G, McDowell A, Black R, Dubochet J (1984) On the preparation of cryosections for immunochemistry. J Ultrastruct Res 89: 65–78

2.2.6 Staining Sections

The contrast achieved through osmium during the fixation can be increased considerably by treating sections with uranyl and lead salts (see Fig. 2.16 a–d). These heavy metals are either bound directly or are precipitated, leading to a selective "staining" of the section whereby certain parts of the specimen become darker than their surroundings.

2.2.6.1 *Staining Solutions*

The most frequently used staining solutions are uranyl acetate and lead citrate. The former is made up in either water or alcohol at a concentration ranging from 1% (w/v) to saturation. The stain is usually taken up more quickly and more effectively from alcoholic solutions. It reacts principally with nucleic acids, but proteins also become stained. Lead citrate increases the general contrast of membranes, proteins, nucleic acids and glycogen. Responsible for this are phosphate, carboxyl, and sulfhydryl groups which bind lead cations.

Both lead and uranium are poisonous heavy metals. In addition, uranium is weakly radioactive. This means that care is required when working with these

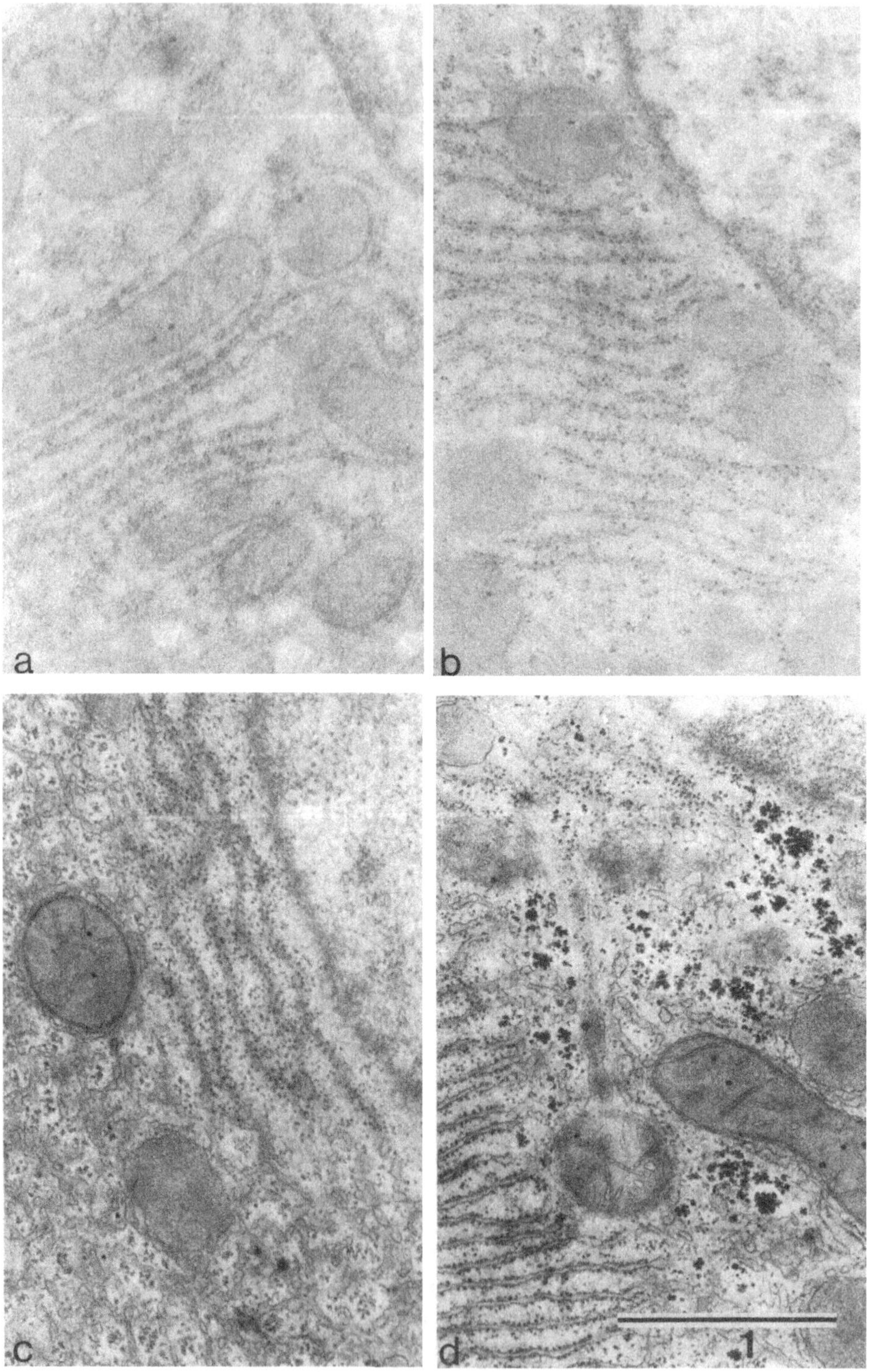

Fig. 2.16 a–d

substances and that arrangements must be made for the proper disposal of waste solutions.

URANYL ACETATE

A. Saturated aqueous solution:

Solid p.a. uranyl acetate (UO_2Ac) is added to 100 ml double-distilled H_2O in a glass-stoppered, brown bottle (or a polyethylene bottle, wrapped in aluminium foil) and the bottle is shaken until the uranyl acetate dissolves. This is done several times over 2-3 h until saturation conditions are reached. The solution is allowed to stand overnight. The supernatant is then carefully pipetted out and either centrifuged (bench centrifuge, 15 min) or passed through a Millipore filter. The final solution has a pH of 4.0 and should be stored in the dark. If the solution becomes turbid it should be discarded.

B. Saturated methanolic solution:

A small amount of uranyl acetate is added to 2-4 ml 70% (w/v) methanol in a centrifuge tube, shaken vigorously (vortex mixer) for a few minutes and then centrifuged to remove undissolved salt. The supernatant is carefully pipetted into a brown glass bottle. The solution is stable for about a week at $-4°$ C and has a pH of 3.5-4.0.

LEAD CITRATE

A. After Reynolds (1963):

1.33 g lead citrate [$Pb(NO_3)_2$] and 1.76 g sodium citrate [$Na_3\ (C_6H_5O) \cdot 2H_2O$] are dissolved in 30-ml double distilled H_2O in a 50-ml ground-glass stoppered bottle or a polyethylene bottle. The solution is shaken vigorously for 1 min and then at intervals over the next 30 min. 8 ml 1 M NaOH (free of Na_2CO_3) is then added together with water to make a volume of 50 ml. The solution is clear and has a pH of 12.

B. After Venable and Coggeshall (1965):

0.01 g lead citrate is dissolved in 10 ml double-distilled CO_2 free water to which 0.1 ml 10 M NaOH has been added. Rigorous shaking (vortex mixer) in a centrifuge tube is often necessary. The solution can be used immediately; it is stable for several weeks when stored in the absence of air to prevent lead car-

◁

Fig. 2.16 a-d. Double staining as demonstrated on sections of rat liver cells (fixed by perfusing with 2.5%, v/v, phosphate buffered glutaraldehyde, post-fixed with buffered 2%, w/v, OsO_4). Sections are of the same thickness and were exposed at the same primary magnification and accelerating voltage (60 kV). **a** Unstained section; **b** after staining with uranyl acetate for 15 min; **c** after staining with lead citrate for 5 min; **d** after staining with both uranyl acetate (15 min) and lead citrate (5 min). Magnification bar in μm

bonate formation (this can be achieved by pressing the air out of a polyethylene bottle before screwing down the cap).

2.2.6.2 Procedure for Double Staining Sections

Double staining of sections from conventionally embedded material with uranyl acetate and lead citrate is a routine procedure in TEM. Sections are usually stained on the grids which were used to pick them up from the trough liquid. However, free-floating sections can also be stained with the same solutions. In this case the sections are transferred from the trough liquid to a watch glass containing the staining solution (see Fig. 2.17b) with the help of a platinum

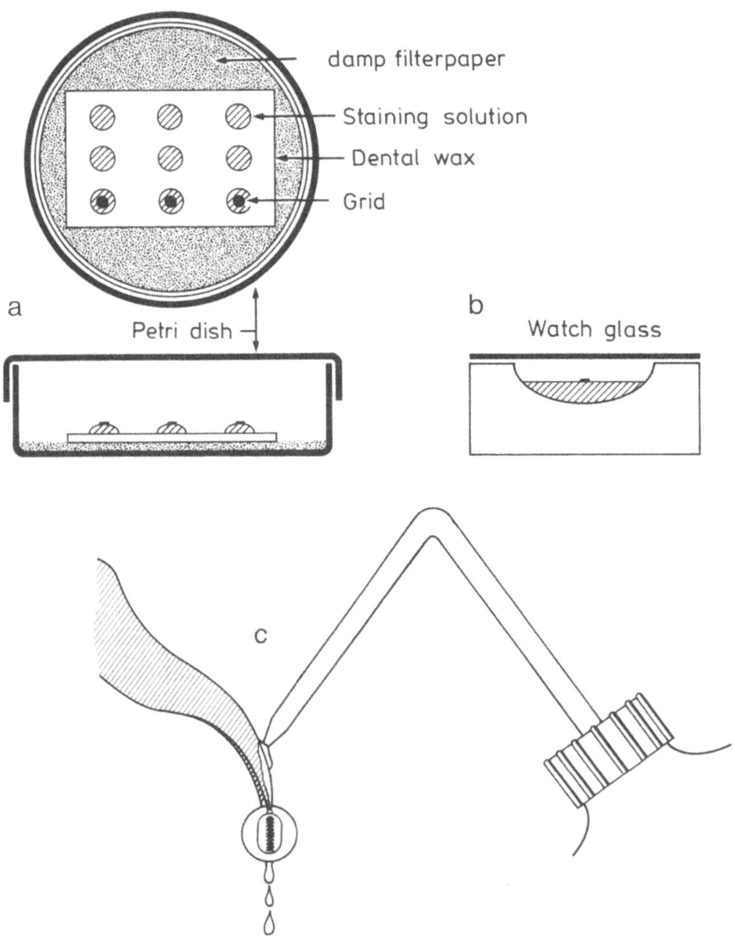

Fig. 2.17 a–c. Staining of sections. **a** in a Petri dish on drops of staining solution (sections on grids); **b** in a watch glass (free-swimming sections); **c** washing sections after staining

loop (or with small plastic rings which one can press out of negative film with cork borers). The subsequent washing steps are also carried out in this manner.

The staining of sections adhering to grids is best carried out in a glass Petri dish fitted with moist filter paper and a piece of dental wax or parafilm (see Fig. 2.17 a). When staining with lead citrate it may be necessary to place one or two moistened pellets of KOH or NaOH on the wax in order to maintain a CO_2-free atmosphere.

Usually one stains first with uranyl acetate. Using a clean Pasteur pipette, the stain is taken from roughly the middle of the stock bottle. The first few drops are discarded and a row of three to five drops is made on the wax. The grids are then placed on the drops of uranyl acetate with the section side downwards. The Petri dish is closed for the duration of the staining.

The grids are then washed free of excess stain by repeated dipping in a series (usually four) of 50-ml beakers filled with double-distilled water. Excess water is removed from the surface of the grids with filter paper before placing them onto lead citrate. Lead citrate staining is carried out in a Petri dish separate from that of the uranyl acetate staining. After the lead staining is completed the grids are washed by holding them in a stream of 0.02 M NaOH from a wash bottle for a second or two, followed by double-distilled water (at least 20 ml per grid) (see Fig. 2.17 c). The grids are finally dried with filter paper (see Fig. 2.14 c) and stored with the section side uppermost on filter paper in a covered Petri dish.

Optimal staining periods have to be worked out for each individual object; they are usually in the range of 5 min for uranyl acetate and lie between 2 and 10 min for lead citrate. Poor staining often results from a precipitation of the heavy metal (as long crystalline needles in the case of uranyl acetate; as fine particles for lead citrate). These precipitates can form during the staining period but may already be present in the staining solution of the stock bottle.

2.2.6.3 Staining Sections of Material Embedded for Immunocytochemical Purposes

2.2.6.3.1 Staining of Glycol-Methacrylate Sections. These are usually stained with a 4% (w/v) uranyl acetate solution in 75% (v/v) methanol for 5–10 min. The staining technique is the same as that just described. The sections are then washed in a descending series of methanol concentrations (75%, 50%, 25%, 5%) before drying on filter paper. Depending on the nature of the material involved, the sections can be additionally stained with lead citrate (also as above).

2.2.6.3.2 Staining of Lowicryl Sections. Both aqueous and saturated methanolic uranyl acetate solutions as well as Reynolds lead citrate can be used for sections of Lowicryl embedded material. As seen in Table 2.10, the staining times with uranyl acetate are usually longer than those regularly employed for conventionally embedded material, especially in the case of HM 20. Between the uranyl

Table 2.10. Staining times for sections cut from Lowicryl-embedded objects

Stain	HM 20	K4M
Uranyl acetate (saturated aqueous solution)	35 min	5–10 min
Reynold's lead citrate	1–3 min	1–3 min

and lead staining the sections are washed free of excess uranyl acetate as described above.

2.2.6.3.3 Staining of London Resin Sections. The staining of sections from LR-white and LR-gold embedded material is in principle the same as for Lowicryl embedded objects. Individual differences in staining periods may, however, exist.

2.2.6.4 Staining Cryosections

In order to impart contrast, frozen thin sections are generally stained with uranyl acetate. A procedure for this is presented in Fig. 2.18. The following solutions are necessary:

a) Uranyl acetate-oxalate pH 7.0 (made by mixing equal volumes of 0.3 M oxalic acid with 49% aqueous uranyl acetate and adjusting to pH 7.5 with 5% (w/v) NH_4OH with the help of pH paper, but not an electrode).

b) Methyl cellulose (Methocel, Fluka). A 2% (w/v) solution of 25 centipoise is made by adding the powder to water preheated to 95° C. After mixing with a magnetic stirrer at 95° C for a few minutes, the solution is cooled in an ice bath and stirred for a further 4–8 h at 4° C. The solution is then allowed to stand for a further 3–4 days at 4° C before centrifuging at high speed, also at 4° C. The centrifuge tubes are then stored in a refrigerator for at least 3–4 weeks without disturbing the pellet. For each experiment small amounts of the solution are carefully removed from the surface layer.

The grid is first washed on drops of distilled water (4 × 1 min) before transferring to the uranyl acetate solution for 5 min. After being washed again on drops of distilled water (3 × 5 min), the grid is allowed to touch, in sequence, two drops of a uranyl acetate-Methocel solution (Methocel mixed with 2–4% aqueous uranyl acetate to give a final uranyl acetate concentration of 0.1 to 0.4%) before allowing to rest on a drop of this solution on ice for 10 min. The grid is then lifted off with a platinum or stainless steel loop (diameter 3.4 mm) and excess liquid removed with filter paper. Keep the grid on the loop for drying, which may take several minutes. A gold-blue interference colour should then become visible. Finally the grid is removed from the loop, firstly by punching several holes in the film surrounding the grid with a fine needle and then by lifting the grid out with fine tweezers. The resulting contrast in the EM may be

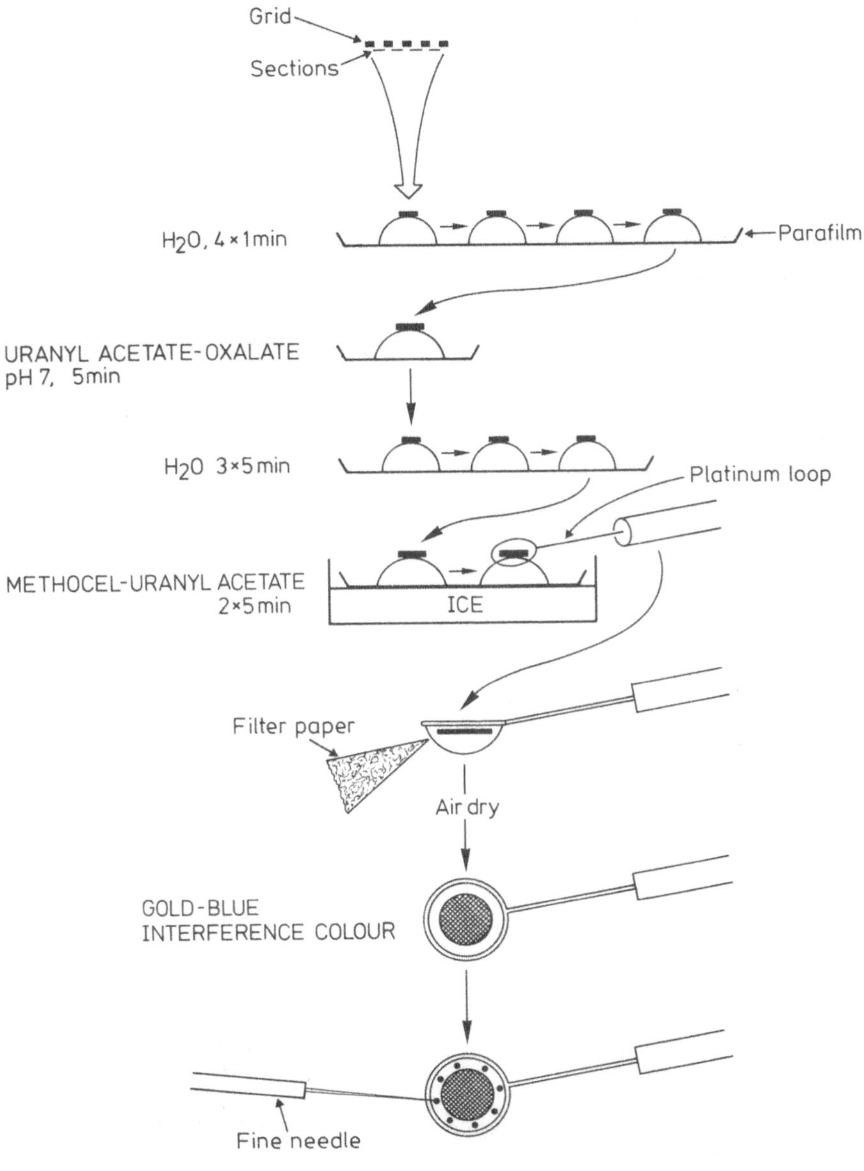

Fig. 2.18. Staining-drying procedure for cryosectioning. This diagram is modefied from information presented in seminars on cryo-ultramicrotomy by Dr. G. Griffith (EMBL, Heidelberg, FRG) and Dr. J. W. Slot (Dept. Cell Biology, Univ. Utrecht, Netherlands)

lower or higher depending upon how much, or how quickly the uranyl acetate-Methocel solution has been removed with filter paper. This step must be varied accordingly.

2.2.6.5 *Block Staining*

Block staining means that the specimens are exposed to heavy metal salt solutions before they become embedded and sectioned. There are two advantages in doing this: one is that the disadvantages of section staining (contamination!) are avoided. More important, however, is the "fixing" effect which this procedure seems to produce on membranes (cf. Fig. 2.19 a, b).

Most often uranyl acetate is used, but lead salts can also be applied in this way. For immunocytochemistry, however, uranyl acetate should be replaced by the same concentrations of phosphotungstic acid. The solutions are usually applied after the wash period following the secondary fixation (note: if a phosphate buffer has been used, wash twice with water before adding the staining

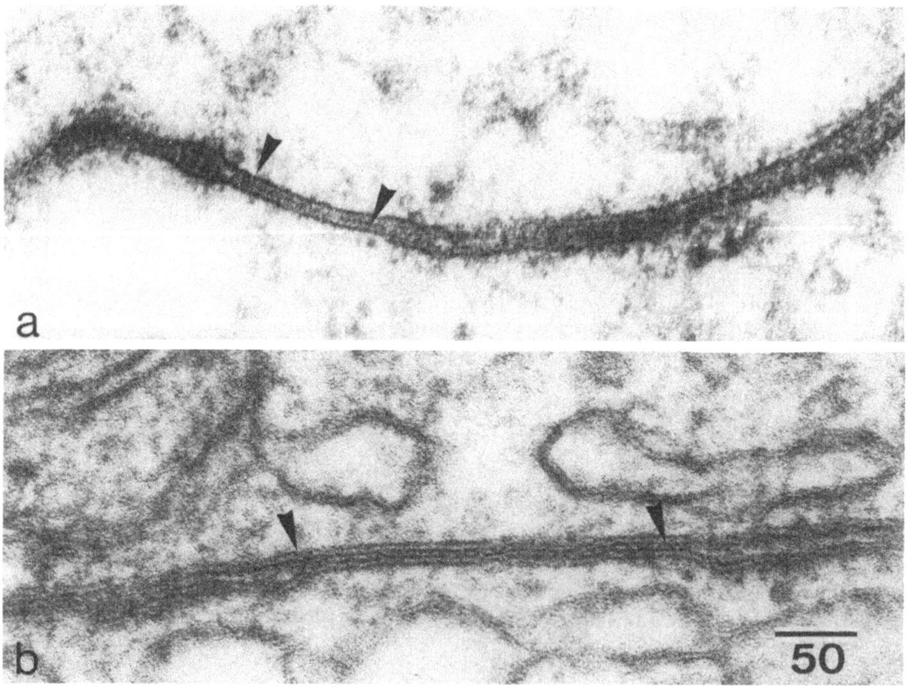

Fig. 2.19 a, b. Plasma membranes and gap junctions *(arrowheads)* in nerve cells of a locust (fixed by immersion in 2.5%, v/v, glutaraldehyde and 2%, v/v, formaldehyde buffered with cacodylate). **a** Section double-stained with uranyl acetate (15 min) and lead citrate (5 min); **b** block staining with uranyl acetate and poststaining with lead citrate. Magnification bar in nm. (Original micrographs F. Killmann)

solutions to avoid stain precipitation) but can also be applied during the dehydration, e.g. with methanolic (70%) uranyl acetate.

A well-known method of block staining is that of Karnovsky (1967):

a) *Required Solutions:* dissolve 23.3 g maleic acid in 1 l distilled H_2O containing 200 ml 1 M NaOH. A 0.05 N buffer solution is obtained by adding 14 ml (for pH 5.2) or 54 ml (for pH 6.0) 0.1 M NaOH to 50 ml of the maleic acid solution and adding H_2O to 200 ml. A 0.5% uranyl acetate solution is made up with this buffer. Turbidity of the solution is not unusual.
b) *Procedure:* wash the fixed material several times in maleic acid buffer for a period of 30 min. Immerse the material in staining solution at 4° C in the dark for 2 h. Proceed directly into the dehydration solvent series.

Literature

Hayat MA (1975) Positive staining for electron microscopy. Van Nostrand-Reinhold, New York, pp 1-361
Karnovsky MJ (1967) The ultrastructural basis of capillary permeability studied with peroxidase as a tracer. J Cell Biol 35: 213-236
Reynolds ES (1963) The use of lead citrate at high pH as an electron opaque stain in electron microscopy. J Cell Biol 17: 208-212
Tokuyasu KT (1978) A study of positive staining of ultrathin frozen sections. J Ultrastruct Res 63: 287-307
Venable JH, Coggeshall R (1965) A simplified lead citrate stain for use in electron microscopy. J Cell Biol 25: 407-408

2.3 Macromolecular EM

2.3.1 Isolated Proteins and Protein Aggregates

2.3.1.1 Preparation of Specimens

The aim of electron microscopy of isolated proteins and protein aggregates (both natural, e.g. microtubules or bacterial flagella and artificial, e.g. two-dimensional protein crystals, see Sect. 2.3.1.4) is to photograph the object at the highest resolution possible, and, from the micrograph, to derive an appropriate three-dimensional reconstruction.

Because isolated macromolecules rarely show a specific orientation when lying on a support film, several different views or projections of one and the same particle are possible when investigating such structures in the EM. Thus, it is imperative that preparations of isolated proteins be absolutely pure - unknown, contaminating, proteins should not be present.

A number of agents are often included in the extraction media for proteins. These are necessary for improved extraction from membranes, e.g. detergents, or they help to stabilize the isolated protein, e.g. glycerol. Often the presence of such substances leads to inferior quality micrographs. Hence, whenever possible, detergents, glycerol, sugars etc. should be removed from the purified protein by dialysis before preparing specimens for electron microscopy. Even when salts are present their concentration should be reduced to a physiological minimum.

The concentration of protein in a specimen for EM investigation should lie between 20 and 200 μg ml^{-1} in order to achieve an adequate density of particles on the support film. Should other concentrations be necessary, e.g. in the case of protein aggregates concentrations of 1 mg ml^{-1} are not unusual, the appropriate concentration can only be determined by preparing a concentration series.

2.3.1.2 Negative Staining Techniques

Negative staining is not just a method restricted to the visualization of proteins (Fig. 2.20a–d); it can also be used for lipoproteins, nucleoprotein complexes, fibrillar polysaccharides, for isolated organelles, e.g. mitochondria or fragments thereof, as a quick method for checking the quality of fractions during the isolation of a particular cell component (Fig. 2.20e, f) as well as for whole cells, e.g. bacteria (Fig. 2.20g, h). The principle is the same in all cases, namely the application of a heavy metal salt solution which dries down around the specimen on the support film. As a result the specimen itself remains electron-translucent whilst its immediate surroundings do not. However, a positive staining of some

Fig. 2.20. a, b Bacterial glutamine synthetase negatively stained with uranyl acetate. The preparations differ in the relative amounts of staining solution which have been sucked off the grid during the final drying step (**a** little stain remaining, "shallow stain"; **b** a lot of stain remaining, "deep stain"). In both micrographs the subunit structure of individual molecules is visible; the enzyme particles lie with differing orientations with respect to the electron beam; thus, different projections of the molecules are present in the final image. Magnification bar in nm.
c Isolated citrate lyase negatively stained with uranyl acetate. The enzyme is present in "ring" and "star" forms (see also Fig. 2.27k and Fig. 4.1a). Both forms are composed of six large, six medium and six small subunits, set together in two layers. Magnification bar in nm.
d Isolated BglI-restriction endonuclease molecules negatively stained with uranyl acetate. *1–7* the diameter of the complexes is less than that in *8* because they are supported by much more dried staining salt ("deep stain"). In *8* the stain was extensively sucked off the grid during drying ("shallow stain"). The micrographs show various projections of one and the same particle complex. In *1–4* and *8* the subunit structure is clearly visible. (After Johannssen et al. 1979).
e, f The bacterium *Rhodospirillum rubrum* negatively stained with uranyl acetate. **e** portion of a flattened cell with internal photosynthetic membranes; **f** isolated photosynthetic membranes, contaminated with flagellar fragments. Magnification bar in μm.
g The pole of a negatively stained (phosphotungstic acid) *Escherichia coli* cell showing type-I pili (see also Fig. 2.21a, c). Magnification bar in nm. **h** Negatively stained (uranyl acetate) bacterium *(Alcaligenes eutrophus)* with peritrich flagella *(arrowheads)*. Magnification bar in μm

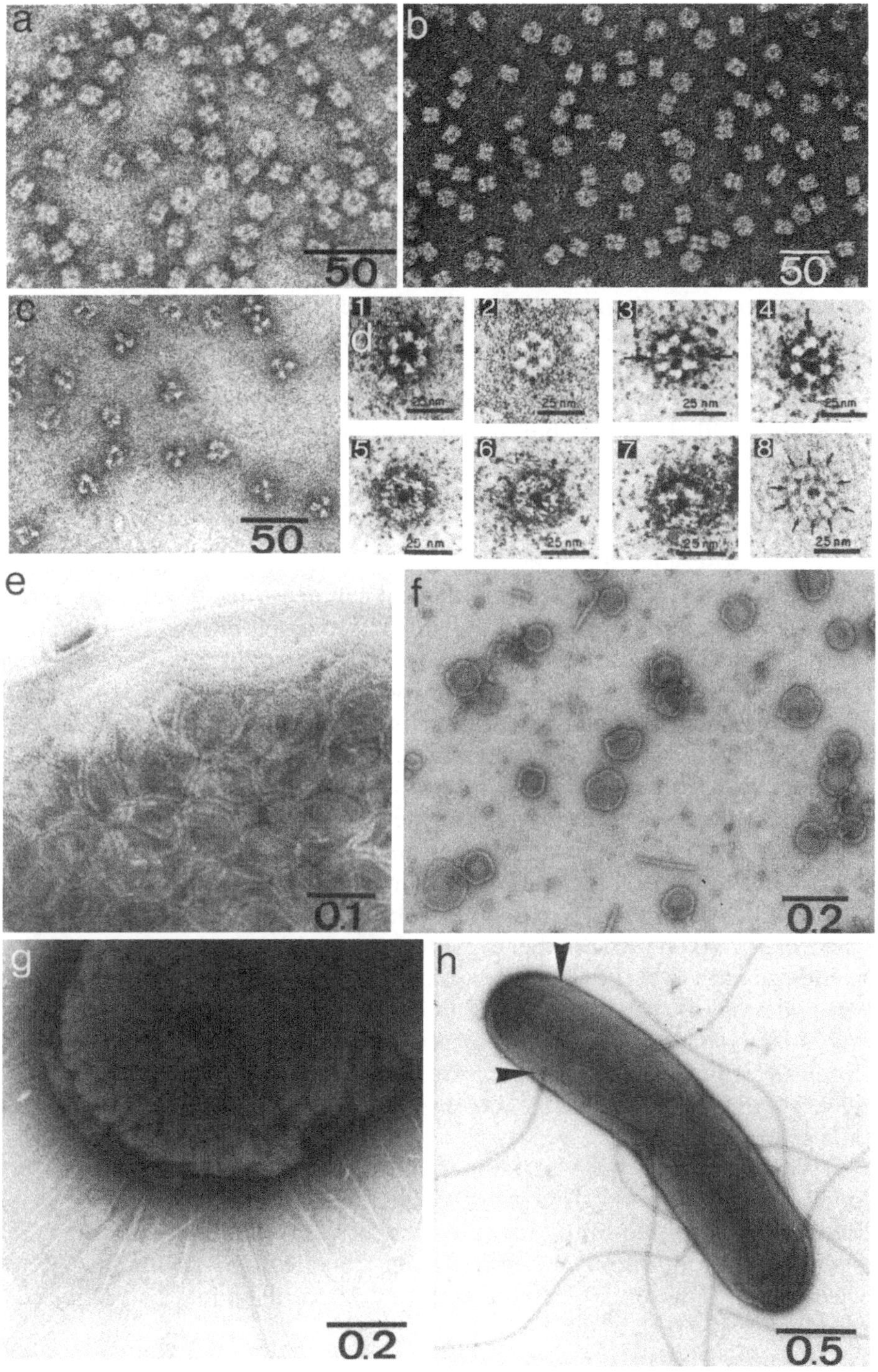
a
50
b
50
c
50
1 d
25 nm
2
25 nm
3
25 nm
4
25 nm
5
25 nm
6
25 nm
7
25 nm
8
25 nm
e
0.1
f
0.2
g
0.2
h
0.5

Fig. 2.20 a-h

components, depending on the nature of the specimen and the staining solution, is always possible. Moreover, fine details can always be occluded due to staining solutions penetrating into hollow spaces within the specimen, or they can be masked by a thick overlying layer of stain. Typical negative staining solutions are uranyl acetate (1-4%, w/v, dissolved in double distilled water - the pH lies around 4.7-4.9), phosphotungstic acid (1-2%, w/v, pH brought to 7 with NaOH or KOH) and uranyl formate (made up as for uranyl acetate but must be used immediately upon preparing).

In order to avoid deformation or dissociation of the specimen during the drying step, it is often necessary to stabilize the particles before staining by quickly (several seconds to several minutes) fixing in a buffered glutaraldehyde solution. If the specimens contain high (0.2-2%, v/v) salt concentrations, or sugars or glycerol are present, it is advisable to include one or two washing steps (in distilled water or buffer) before applying the negative stain.

Negatively stained specimens can be inserted in the EM within minutes after their preparation. They are relatively stable and can be stored in a desiccator for several weeks.

When observing negatively stained specimens in the EM, low dose conditions are to be recommended in order to keep heavy metal salt crystallisation, and the resulting loss of resolution (which usually lies around 2 nm), to a minimum. Whenever possible a cold stage should be employed. One should note here that the final image in this type of specimen preparation, as with some others, represents a combination of all the individual superimposed structural details present at different levels in the specimen (see also Chap. 4.2.3).

Specimens which contain particles of different mass, size or composition (e.g. intact enzymes plus enzyme fragments or mitochondrial membranes and dissociated ATPase fragments) have to be mounted on the support film with the help of an atomizer (spray gun). In so doing, one is aiming for a quantitative and stable adhesion of the particles on the film. To this end the particle suspension is normally mixed with the staining solution and the two are sprayed together. It is important here that the staining solution does not give rise to turbidity or to precipitation of components out of the particle suspension - one must therefore avoid large changes in pH or in salt concentration. Uranyl acetate and uranyl formate with pH values of under 5 are therefore unsuitable for this type of negative staining. Instead one should employ 0.5-1%, w/v, phosphotungstic acid at a pH of 7.

With this method it is easy for small particles to become obscured by too much staining solution. This happens when the sprayed droplets are too large, and can be corrected for by using a finer spray or by diluting the heavy metal staining solution. Improved results can be obtained by addition of glycerol to the spraying solution. The concentration of glycerol should be optimized, depending on the type of sample.

A second method for negative staining simply entails pipetting a drop of the particle suspension onto a support film. As with the spray method, it is better to mix the particle suspension with the staining solution before applying to the

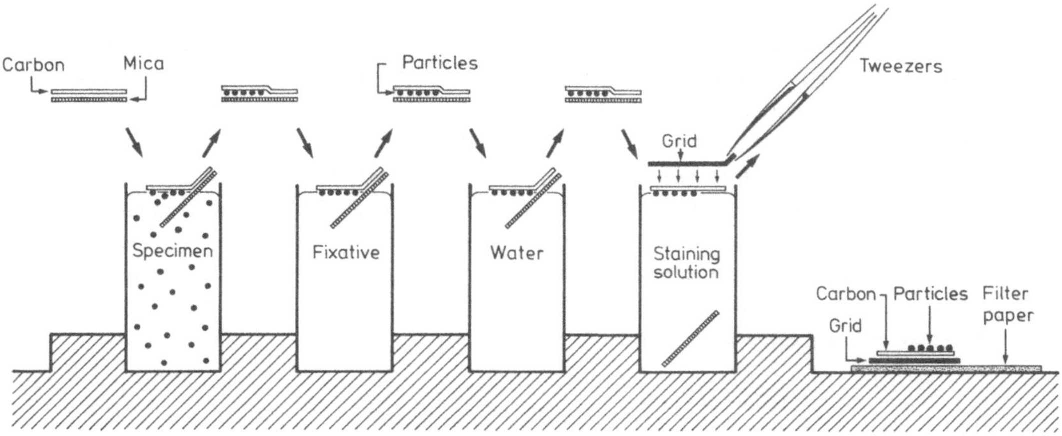

Fig. 2.21. Experimental set-up for negative staining

film. In this way the possibilities of drying artifacts occurring are reduced by half. Moreover, if the particle suspension is applied to the film first and the excess liquid removed immediately with filter paper before adding the staining solution, there is a very good chance that many of the smaller particles will not have become attached to the film. This method should therefore not be used for quantitative purposes.

Both of the above methods have one cardinal disadvantage: the salt concentration increases dramatically during the latter stages of drying down the specimen onto the film. This can lead to considerable structural damage. The heavy metal portion of the salt is unavoidable but the presence of salts from the particle suspension can be overcome thanks to a technique introduced by Valentine et al. in 1968 (see Fig. 2.21 for experimental set-up). In this method the particles are allowed to adsorb onto the surface of a carbon support film previously deposited onto a piece of cleaved mica as described in Chapter 1.4.6. In order to do this one cuts a 3 × 3 mm piece of carbon-coated mica and inserts it at an angle into the particle suspension. The carbon film floats off the mica but by holding the mica with fine tweezers it still remains attached at one corner. When the mica is removed from the suspension the carbon film is brought with it, sandwiching the particles between mica and carbon. The density of particles which are adsorbed onto the carbon film is determined by their concentration in the specimen, their mass and by the length of time in which the carbon film is allowed to float on the suspension. "Floating times" of 10–60 s are usual. Since the particles remain attached to the carbon film, a series of fixation and washing steps can easily be carried out by simply inserting the mica into each solution, briefly floating the particle-covered carbon film and withdrawing as above. Finally, one floats the carbon film onto the negative staining solution and releases the mica, which sinks. The carbon film is then picked up from above or from below with a grid and dried on filter paper. One should avoid sucking off too much stain with the filter paper since sufficient stain must remain both for reasons of contrast and also to act as an aid in supporting the particles on the carbon film. If the carbon film is picked up from below it is possible to use a carbon or collodion-coated grid. The particles then become "sandwiched" between the two films together with stain. Careful drying with filter paper ensures that sufficient stain will remain to prevent a deformation of the two films. Although this third method of negative staining is clearly the most reliable and reproducible, it has certain disadvantages when a specimen contains particles of different sizes and properties. The first step is diffusion-dependent and can lead to a proportional over-representation of smaller, as opposed to larger particles on the carbon film.

Literature

Johannssen W, Schütte H, Mayer F, Mayer H (1979) Quaternary structure of the isolated restriction endonuclease Endo R. BglI from *Bacillus globigii* as revealed by electron microscopy. J Mol Biol 134: 707-726

Mayer F, Spiess E (1976) Negativkontrastierung (Film C 1191) Wissenschaftlicher Lehrfilm des IWF Göttingen (Institut für den wissenschaftlichen Film, Nonnenstieg, D-3400 Göttingen) aus der Reihe Elektronenmikroskopische Präparationsmethoden (also available in English)

Valentine RC, Shapiro BM, Stadtman ER (1968) Regulation of glutamine synthetase. XII. Electron microscopy of the enzyme from *Escherichia coli*. Biochemistry 7: 2143-2152

2.3.1.3 High Resolution Metal Shadowing

This is a method which is not only useful for depicting isolated macromolecules (not, however, nucleic acids), but has also been successfully employed on objects as large as whole cells (e.g. Fig. 2.22 a-c). The metals most usually employed in this technique are platinum, iridium, platinum-iridium mixtures, tungsten or tantalum. Contrast is achieved by depositing particles of these metals at an angle of between 30° and 45°. As a result, one side of the object becomes coated with a metal film, the other relatively less so or not at all. In the EM the regions which lie in the "shadow" thus appear brighter than those which bear the heavy metal.

Resolution in heavy metal shadowing depends on the metal in question and the type of evaporation technique. The most common evaporation method is that of resistance-heating. Here platinum-iridium wire (80:20, w/w; 0.1 mm thick; length according to predetermined testing) is wound around a piece of thick tungsten wire which connects the two electrodes. A metal film deposited in this way has a resolution of between 4 and 3 nm. This can be improved upon when the metal to be evaporated is contained in a carbon holder. During the heating some of the carbon sublimes and condenses on the surface of the object together with the metal (see Fig. 2.22 a, b). The presence of carbon apparently

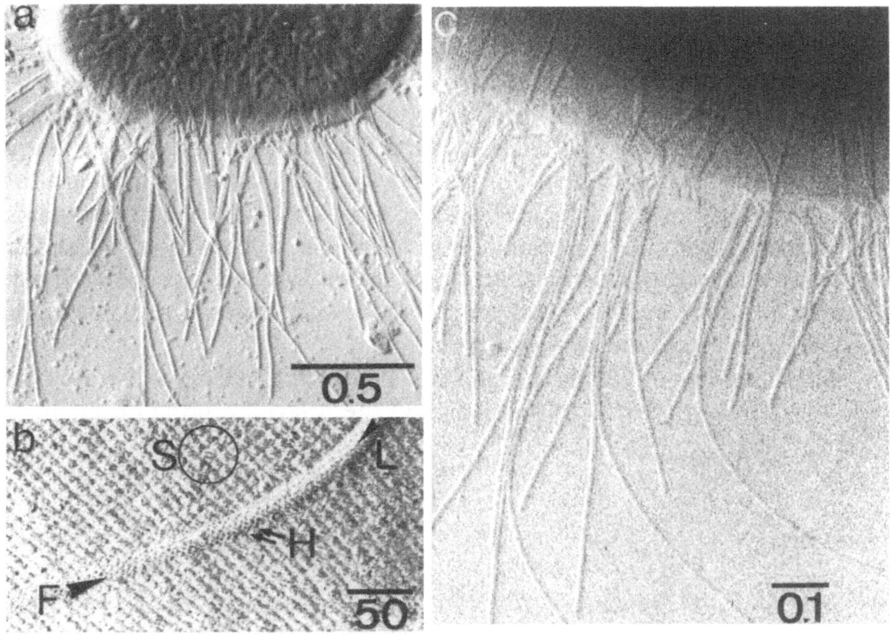

Fig. 2.22 a–c. Heavy metal shadowed preparation. **a** Conventional platinum-iridium-carbon mixed evaporation (thermal resistance evaporator). *Escherichia coli* cell with type-I pili. **b** Conventional platinum-iridium-carbon evaporation (prepared as a freeze-etch replica). Surface of the bacterium *Clostridium aceticum* (Braun et al. 1981). *S* surface layer with a periodic substructure; *F* flagella with ultrastructural details; *H* helically arranged flagellin molecules; *L* flagellin molecules present in long rows. **c** Object as in **a**, but shadowed with an electron beam evaporator (platinum-carbon). Magnification bars in **a** and **c** in μm; in **b** in nm

prevents the aggregation of metal particles which can occur when the object is subjected to a beam of electrons in the EM.

Another way to evaporate heavy metals (and carbon as well) is by using an electron beam gun (see also Sect. 2.6.5). Here the metal which is to be evaporated serves as the anode and is heated by electrons which impinge upon it, from a spiral-formed cathode. Since the metal particles are usually smaller and more homogeneous than those obtained by conventional resistance evaporation, the resolution of metal films deposited by an electron beam gun is of the order of 2 nm. In practice these evaporations are much less reproducible than those of platinum-carbon. Moreover, tungsten-tantalum films are much less stable – they lose contrast when not stored in an exsiccator.

The preparation of specimens for heavy metal shadowing is more or less identical to that already described under Sections 2.3.1.1 and 2.3.1.2. When specimens are to be sprayed onto grids glycerol (optimal concentration varies from object to object) is used and the negative staining solution is omitted. Most important, however, is that specimens should be washed well before shadowing, since dried-down salts, sugars and other contaminants can often prevent good visualization of the particles in the specimen. One must nevertheless realize that drying artifacts are much more of a problem in the preparation of metal-shadowed objects since a protective coat of staining reagent, as in negatively stained preparations, is absent.

In contrast to negative staining only the upper surface of the specimens is made visible after heavy metal shadowing. Estimates of particle size are possible by measuring the length of the shadow and calculating from the shadowing angle the height of the particle. However, for many reasons such estimates can differ greatly from the true diameter of a particle. For example, the shadow is sometimes not clearly delineated; the support film may not be perfectly flat (leading to a false angle of shadowing); shadowing material may pile up on the side of the particle, adding to its true height.

Modifications of the unilateral metal shadowing technique are the bilateral (two evaporations at 90° to another) and the rotary evaporation method (see Sect. 2.3.2.2).

Literature

Braun M, Mayer F, Gottschalk G (1981) *Clostridium aceticum* (Wieringa), a microorganism producing acetic acid from molecular hydrogen and carbon dioxide. Arch Microbiol 128: 288-293

Spiess E, Mayer F (1976) Kontrastierung durch Metall-Schrägbedampfung (Film C 1192) Wissenschaftlicher Lehrfilm des IWF Göttingen (Institut für den wissenschaftlichen Film, Nonnenstieg, D-3400 Göttingen) aus der Reihe Elektronenmikroskopische Präparationsmethoden (also available in English)

2.3.1.4 Preparation and Imaging of Two-Dimensional Protein Crystals

Particle deformation is a problem which is inherent in the electron microscopy of proteins (see Sects. 2.3.1.2 and 2.3.1.3). Particularly susceptible in this regard are individual, flat-lying macromolecules such as enzyme complexes. This problem can be overcome to a certain extent when the molecules are brought together as an ordered two-dimensional aggregate or crystal on the support film (see Fig. 2.23 a-c). Deformation is then reduced due to the "supporting" effect of neighbouring particles.

Protein crystals of this type are prepared through microdialysis whereby a purified protein preparation (in active form when it is an enzyme) of concentration around 1 mg ml^{-1} and volume of 0.1 ml or less can act as the starting solution. For the dialysis a 400 mesh grid with a collodion support film (no holes!, can be strengthened with carbon) is used. A small droplet (diameter 1 mm or less) of the protein solution is added to the middle of the grid with the help of a plastic micropipette tip fitted with a rubber teat. During this procedure the grid is held with fine tweezers without allowing it to come into contact with a surface. Provided the drop retains its form (if not, throw away) on the collodion film the dialysis can be started by carefully placing the grid onto the surface of a

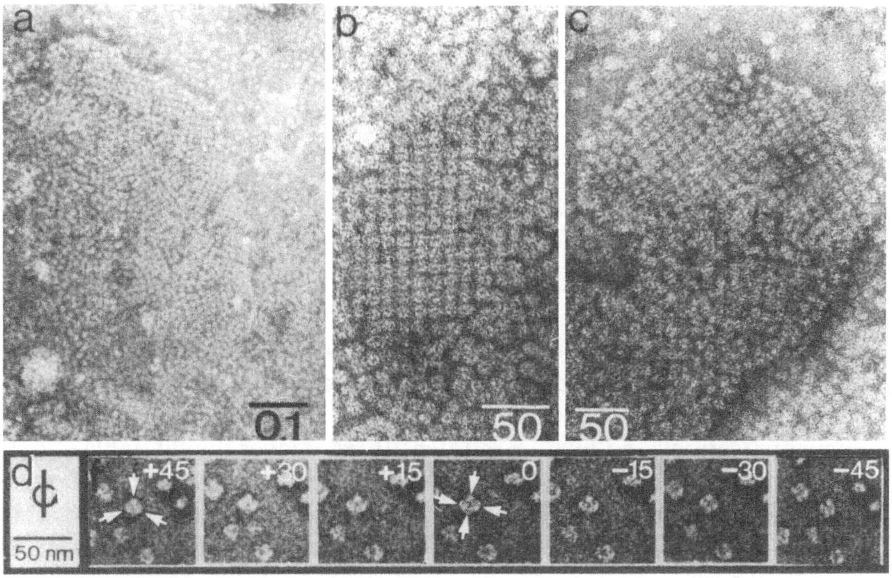

Fig. 2.23. a–c Artificially induced two-dimensional (2-D) crystalline aggregates of the enzyme ribulose bisphosphate carboxylase isolated from the bacterium *Alcaligenes eutrophus*. **a** a group of small 2-D crystals, together with an irregular aggregate (*left* in picture) as well as separately lying individual molecules. **b, c** as for **a**, but at higher magnification. Negative staining with uranyl acetate. Magnification bar in **a** in μm; in **b** and **c** in nm.
d Tilting series of a specimen of the enzyme pyruvate carboxylase isolated from the fungus *Aspergillus nidulans* (tilting angle and axis of tilt are given). Depending on the angle of tilt the enzyme complex can take on a tetrahedral (+45°) or rhomboid form. The subunits are indicated with *arrows*. Negative staining with uranyl acetate (Osmani et al. 1984)

buffer solution contained in a small glass or plastic vessel. The appropriate buffer is to be determined beforehand through trial and error involving a systematic testing of different buffer types, concentration and pH.

When correctly performed it is possible that different types of protein crystals may form depending upon the orientation of the first (seed) protein molecule on the support film. Moreover it is possible that noncrystalline aggregates may be present. If, however, three-dimensional crystals are formed one should either reduce the length of the dialysis period or the protein concentration.

As an alternative to negative staining, crystalline protein aggregates can, of course, be metal shadowed. Since these aggregates are periodic structures, special methods of image analysis (see Chap. 4.2.4) can be applied to the negatives obtained, provided that a structure of at least 10 × 10 particles has been imaged.

Literature

Keegstra W, Bruggen EFJ van (1980) A simple way of making a 2-D-array. In: Baumeister W, Vogell W (eds) Electron microscopy at molecular dimensions. Springer, Berlin Heidelberg New York, pp 318–327

2.3.1.5 *Making a "Tilt Series"*

So-called "tilt series" are necessary in order to visualize an object three-dimensionally. They are obtained by tilting the object with respect to the electron beam and exposing at a number of defined angles (up to 60°). For this the EM should be fitted with a goniometric object holder. Simple tilting devices often cause problems with the proper orientation of the sample. The individual micrographs represent projections in particular planes but taken together they enable a spatial reconstruction of the object to be made (see, for example, Fig. 2.23 d).

In preparing a series of tilted exposures there are a number of points which have to be taken into consideration. Firstly since beam damage to the object is dependent upon the beam dose (see Chap. 1.1.1.2), tilt series are best made at low magnifications and low beam intensity. Moreover the tilt angle should be chosen and adjusted at magnifications lower than those used for the exposures. Second, it is imperative that the series be completed as quickly as possible; to this end it may be necessary to uncouple the motorized goniometer and proceed manually. Thirdly, and particularly at high angles of tilt, it is necessary to check and correct for astigmatism and to refocus (even with a eucentric goniometer) before each exposure. Finally one must realize that, at high tilt angles, the image is only in focus along one plane through the object, i.e. that of the tilt axis. Thus those objects lying above and below this axis will be considerably out of focus.

Literature

Osmani SA, Mayer F, Marsten FAO, Selmes IP, Scrutton MC (1984) Pyruvate carboxylase from *Aspergillus nidulans:* Effects of regulatory modifiers on the structure of the enzyme. Eur J Biochem 139: 509–518

2.3.2 Isolated Nucleic Acids

2.3.2.1 *Problems and Aims*

Isolated nucleic acids (NAs) are investigated ultrastructurally in order to gain information on (a) their form (linear, covalently closed circular, open circular; Fig. 2.24 a, b), (b) their length, (c) their molecular weight and the number of bases or base pairs, (d) the distribution of AT- or GC-rich regions (of use, for example, in the identification of nucleic acids or regions of the molecule; Fig. 2.24 e, f), (e) the position of certain genes, (f) mutants through the comparative analysis of similar base sequences in different NAs, (g) evolutionary aspects of NA structure.

A precise and reproducible measurement of the NA in question, irrespective of whether the specimens are double- or single-stranded DNA, RNA or portions of molecules, is a necessary prerequisite in all of these investigations. Because of artifactual changes in form (NAs are very much more long than they are broad) which may occur as a result of the preparation, conventional methods, such as metal shadowing and negative staining (as described above) cannot be employed for NAs. Instead one uses the so-called Kleinschmidt technique or one of the many variations of this method (see Sect. 2.3.2.3). The principle of this method is based on the fact that nucleic acids which have been adsorbed onto a monomolecular proteinaceous supporting film (e.g. of cytochrome c) are quite stable and are protected from mechanical damage. Cytochrome c not only forms a film but also binds to the NA molecules, forming NA-protein complexes whose thickness is much greater than that of the NA itself, making it more recognizable in the EM. Moreover, being embedded in a monolayer imposes a two-dimensional constraint on the three-dimensional form of the NA, making it easier to measure.

After transferring the protein film to an appropriate grid the NA molecules are stained (Fig. 2.24i). To improve the attachment of the protein film to the support film (usually collodion, without carbon backing) the support film should be hydrophilized as in Chapter 1.4.4. When metal shadowing is employed the carbon which is deposited together with the metal is sufficient to stabilize the film. If, however, a method is used which does not involve cytochrome c it is advisable to employ carbon-coated collodion grids.

Literature

Brack CH (1981) DNA electron microscopy. CRC-Crit Rev Biochem 10 (2): 113–169

Fisher HW, Williams RC (1979) Electron microscopic visualization of nucleic acids and of their complexes with proteins. Annu Rev Biochem 48: 649–679

Kleinschmidt A, Zahn RK (1959) Über Desoxyribonucleinsäure-Moleküle in Protein-Mischfilmen. Z Naturforsch 14b: 770–779

Mayer F, Spiess E (1976) Spreitungstechnik zur Präparation isolierter DNA (Film C 1193) Wissenschaftlicher Lehrfilm des IWF Göttingen (Institut für den wissenschaftlichen Film, Nonnenstieg, D-3400 Göttingen) aus der Reihe Elektronenmikroskopische Präparationsmethoden (also available in English)

2.3.2.2 Specimen Preparation

The NA preparation for EM investigation should be checked for purity, through, for example, agarose gelelectrophoresis. It should not contain any other type of macromolecule, although small amounts (end concentration 10 μg ml^{-1}) of proteolytic enzymes, e.g. proteinase K can be tolerated should the usual phenol treatment for removal of proteins prove to be deleterious to the NA.

For the evaluation of the results obtained it is important that the standards are also subject to the identical preparative procedures. Although it is known that, under constant conditions (ionic strength, pH, temperature, cyt c concen-

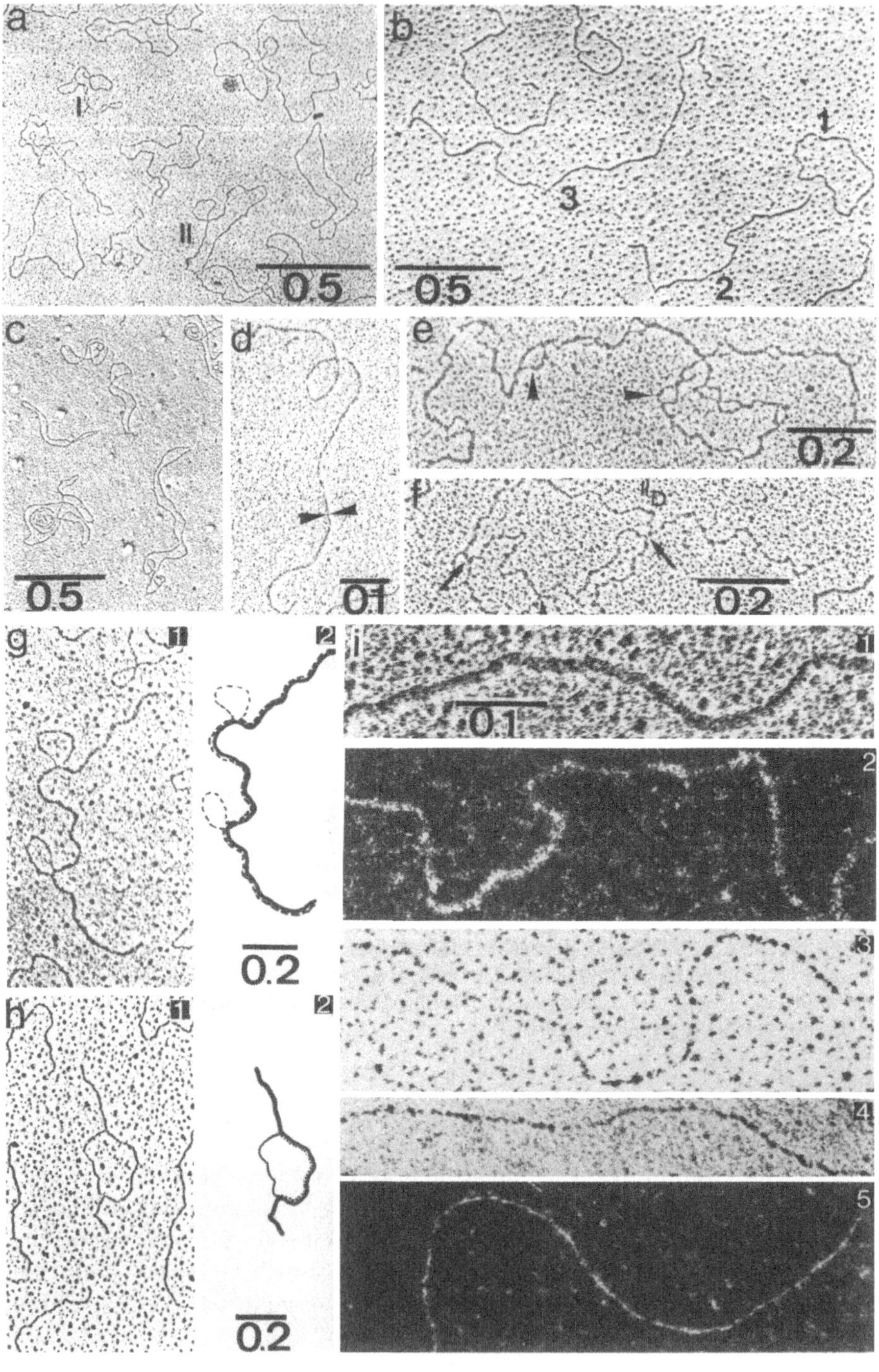

Fig. 2.24 a–i

tration, concentration of formaldehyde or formamide) a linear relationship between NA length and molecular weight exists, it is always better to prepare a specimen of a NA of known length (determined by sequencing) and of similar nature (linear, circular). An external standard allows an approximate size correlation to be made. More accuracy is obtained when the unknown specimen and the standard are mixed together. Of course, one must be careful not to choose an internal standard which resembles too closely the specimen under investigation.

Normally the concentration of the NA to be investigated should be in the region of several μg ml^{-1}. All solutions, except those containing NA, should be filtered (0.2 μm) immediately before use.

2.3.2.3 Spreading and Diffusion Techniques Which Employ Cytochrome c

a) Cytochrome c Spreading Methods Without Formamide

Ammonium acetate (end concentration 0.2 M in dist. H_2O containing 1 mM EDTA), cytochrome c (50 μg ml^{-1} end concentration, diluted tenfold from a fil-

◁

Fig. 2.24. a Double-stranded DNA molecules isolated from the virus SV 40 and prepared with the cytochrome c spreading technique. *I* Covalently closed circular molecule (ccc); *II* "open" circular molecule (oc). Conventional rotary shadowing (Mayer et al. 1975).
b DNA from plasmid pBR 322 partially digested with the restriction enzyme SalGI; cytochrome c spreading technique. *1* monomeric, "open" circular form; *2* monomeric linear form; *3* dimeric linear form. Conventional rotary shadowing with platinum-iridium. (Original micrographs W. Johannssen).
c, d Spreading with BAC (see text); rotary shadowing with an electron beam source; **c** plasmid pGW10; induced molecular deformation through bad spreading; **d** plasmid pGW10 DNA made linear through SalGI treatment; measured fibre width 6 nm. Background is much less "grainy" than in **b**. (Original micrographs W. Johannssen).
e, f Partially denatured double-stranded DNA; cytochrome c spreading; conventional rotary shadowing (Mayer et al. 1975). **e** Segment of DNA isolated from a bacteriophage. *Arrowheads* indicate where the strands have opened up. **f** Strongly denatured circular DNA isolated from virus SV 40. *Arrows* indicate positions where the DNA is still double-stranded.
g, h Heteroduplex structures; cytochrome c spreading; conventional rotary shadowing. (Original micrographs H. J. Lünsdorf). **g** DNA-DNA heteroduplex prepared from mitochondrial DNA from the fungi *Aspergillus nidulans* and *Aspergillus nidulans* var. *echinulatus.* The DNA was made linear with a restriction endonuclease. Single-strand loops are recognizable on the basis of their reduced thickness and contrast (compared the double-stranded DNA). *1* EM micrograph; *2* accompanying line drawing. **h** R-loop (i.e. mRNA-DNA heteroduplex) present on a segment of linearised mitochondrial DNA from *Aspergillus nidulans.* The hybridised mRNA in the DNA double-strand is indicated by *double-lines* in the drawing.
i The appearance of double-stranded DNA after different methods of preparation. *1, 2* Cytochrome c spreading; *3-5* BAC spreading. *1-4* Carbon support film of medium thickness (5-18 nm); *5* very thin carbon film stabilised over a holey support film. *1, 3* conventional rotary shadowing; *2* unstained; *4, 5* positive staining with uranyl acetate in acetone. *1, 3, 4* Bright field exposures; *2, 5* dark field exposures. All micrographs are of DNA from the bacteriophage gd and have the same final magnification. (Original micrographs U. Hahn). Magnification bar in μm

tered stock solution), NA (0.3–0.5 µg ml^{-1}) and dist. H_2O are pipetted into a small plastic test tube and carefully mixed. This solution is termed the hyperphase.

The spreading is carried out in a flat vessel, e.g. a small Petri dish. This is filled with an ammonium acetate solution of lower ionic strength (0.01 M, pH 6.8 containing 1 mM EDTA) constituting the "hypophase". The surface of this solution is "cleaned" by drawing a Teflon-coated metal bar across it. A clean, fat-free glass slide is then inserted into the hypophase at an angle (it acts as a "launching ramp") and the surface of the hypophase is slightly sprinkled with talcum powder. The hyperphase is sucked into a pipette and a drop is added to the glass slide; it runs down the slide and onto the surface of the hypophase. When done correctly, a uniform, half-moon-shaped "spreading field" is formed which remains stable for several minutes. This field is clearly visible by the displacement of the talcum.

A grid with a collodion-supporting film is then gently touched to the surface of the spreading field (it does not matter where). The protein film and NA stick to the surface of the collodion film. The drop of liquid which remains attached to the supporting film is removed by first lowering the grid onto dist. H_2O until contact is made, leaving it there for a few seconds and then transferring it to 90–95% (v/v) ethanol for 10–15 s. The grid is then transferred (with the film surface still downwards) to filter paper for drying. A number of grids can be prepared in this way from the same spreading field (from different positions of course) and their processing is best carried out using a microtitre plate.

Spread NA preparations are usually visualized in the EM after metal shadowing. Contrast can, however, be increased by initially positively staining with uranyl acetate. Sometimes positive staining alone is to be recommended. This is the case when one expects to deal with regions of single-stranded NA in the specimen. Positive staining allows a much easier delineation between single- and double-stranded NA than does metal shadowing. Staining is carried out between the washing and dehydration steps and simply involves floating or immersing the grid in a solution of uranyl acetate (10 µM in 90% ethanol-prepared immediately before use from a 1 mM aqueous stock solution containing 50 mM HCl) for 30 s.

Rotary shadowing is the usual form of metal shadowing for spread NA preparations. This is a variation of the method already described under Section 2.3.1.3. It involves attaching the grids onto the object holder with the help of double-sided sticky tape and, because of the very small diameter of NA strands, reducing the angle of shadowing to 5°–7°. The object holder is set into rotation with the help of a motor operated from outside the vacuum recipient and the specimens are then shadowed. The distance of the specimen from the electrodes, the speed of rotation, the current density and duration of the evaporation are parameters which must be determined for each type of specimen. Such specimens are best viewed in the EM at 40–60 kV accelerating voltage and at magnifications between 5000 and 12000 ×.

b) Spreading Methods with Formamide

These are further developments of the technique described under (a). Formamide prevents random interactions between bases, thereby leading to an improved spreading of the NA molecules. In addition, formamide tends to increase the stability of the NAs.

Preliminary preparation:

Clean glass slides with acetone and dry. Dip into 0.2 M aqueous ammonium acetate (⅓ of slide remaining dry) and allow to dry on filter paper. Hyperphase (spreading solution):

- 10 µl buffer (1 M Tris-HCl pH 8.5; 0.1 M EDTA)
- 10 µl aqueous cytochrome c (500 µg ml^{-1})
- 5 µl NA (5 µg ml^{-1}; 0.1 OD_{260nm})
- 25 µl double-distilled H_2O
- 50 µl formamide (pure, recrystallized).

Mix carefully immediately before use.

Hypophase:

The hypophase is double-distilled H_2O. Insert glass slide at an angle as described in Section 2.3.2.3 a and lay its end on the edge of the Petri dish. Although sprinkling talcum powder on the surface of the water is not necessary here, it is advisable to use those areas of the spreading film (prepared as in Sect. 2.3.2.3 a) nearest to the glass slide.

c) "Droplet" Method (Microdiffusion Technique)

This method is most suitable when small amounts (10^{-10}-10^{-11} g) of NA are available, and has derived its name from the fact that small drops (of volume around 0.04 ml) are placed on a nonwetting surface, e.g. parafilm. The droplet solution has the following constituents (to be pipetted in the order given):

- double-distilled H_2O	1.00 – (x + 0.18) ml
- 2 M aqueous ammonium acetate with 0.01 M EDTA, pH 7	0.10 ml
- 0.01% aq. cytochrome c (diluted from stock solution before use)	0.03 ml
- 3.5% (v/v) formaldehyde	0.05 ml
- NA (end concentration 0.1 µg ml^{-1})	x
	1.00 ml

Formamide (see Sect. 2.3.2.3) can be used instead of formaldehyde although a change in concentration may be necessary. Mix carefully and place a drop onto parafilm, cover and leave to stand (without vibrations) for 15 min. A collodion-carbon-filmed grid is then allowed to touch the surface of the drop. The grid with the attached protein-NA film is processed as under (a) above.

Literature

Ferguson J, Davis RW (1978) Quantitative electron microscopy of nucleic acids. In: Köhler JE (ed) Advanced techniques in biological electron microscopy II. Springer, Berlin Heidelberg New York, pp 123–171

Inman RB (1967) Some factors affecting electron microscopic length of DNA. J Mol Biol 25: 209–216

Kleinschmidt AK (1968) Monolayer technique in electron microscopy of nucleic acid molecules. Methods Enzymol 12 B: 361–377

Lang D, Mitani M (1970) Simplified quantitative electron microscopy of biopolymers. Biopolymers 9: 373–379

Wellauer PK, Weber R, Wyler T (1973) Electron microscopical study on the influence of the preparative conditions on contour length and structure of mitochondrial DNA of mouse liver. J Ultrastruct Res 42: 377–393

2.3.2.4 "BAC" Technique

A film of BAC can be used instead of cytochrome c (see Fig. 2.25 a, b). However, in practice, this method is more difficult to perform successfully than conven-

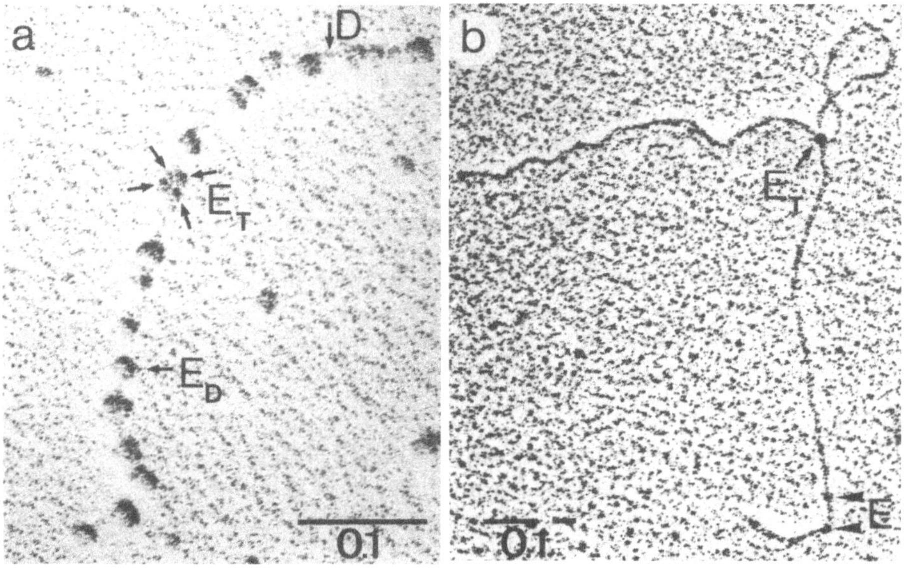

Fig. 2.25 a, b. Double-stranded DNA (from plasmid pGW10 linearised with SalGI)-restriction enzyme EcoRI complexes; BAC spreading; electron beam rotary shadowing. **a** Association of protein and DNA in the presence of Mg^{2+}; a very high binding rate is observed. *D* DNA molecule; E_D dimeric EcoRI (small diameter); E_T tetrameric EcoRI (large diameter). The subunits in the tetrameric (tetrahedral) form are indicated with *arrows*. **b** Association of protein and DNA in the absence of Mg^{2+}; a very much lower binding is observed (see *arrows* E and ET). EcoRI is composed of two dimers, each of which has the same specific binding site. This leads to the production of a DNA loop in **b**. (Original micrographs W. Johannssen). Magnification bar in μm

tional spreading with cytochrome c. The chief disadvantage is "stretching" which can make accurate measurements of NA length impossible.

The BAC technique is usually performed via the droplet method (see Sect. 2.3.2.3 c). The droplet solution (without NA) has the following composition:

- double-distilled H_2O 2.98 ml
- 2 M ammonium acetate with 0.01 M EDTA pH 7.0 15 µl
- BAC-stock solution (20 mg BAC-40%) benzene dimethyl dodecyl ammonium chloride, 60% benzene dimethyl tetradecyl ammonium chloride - dissolved in 10 ml pure, recrystallized formamide). 3 µl

All solutions are centrifuged at high speed (e.g. in a Beckman Airfuge with an A-100 Rotor) to remove large contaminating particles. 0.5-5 µl of a NA solution (2-20 µg ml^{-1}) are added to 100-200 µl of the above solution and mixed carefully. The BAC-NA film is picked up and processed as described under Section 2.3.2.3 c. In order to avoid artificial deformation of NAs grids with hydrophilized support films (see Chap. 1.4.4 for their preparation) are recommended.

Literature

Vollenweider HJ, Sogo JM, Koller T (1975) A routine method for protein-free spreading of double- and singlestranded nucleic acid molecules. Proc Natl Acad Sci USA 72: 83-87

2.3.2.5 *Partial Denaturating, Heteroduplex and R-Loop Techniques*

a) Partial DNA or RNA Double-Strand Denaturation

The basis of this technique is that the DNA is treated in such a way that hydrogen bonds between the base pairs A and T are opened up ("melting"). Under the conditions chosen, the base pairs G and C remain associated with one another. As a result, A-T rich regions of double-stranded NA are made visible as two single strands. In order to perform the "melting" process at low temperature a low ionic strength buffer and dimethyl sulfoxide or formamide must be present. With this method one must take precautions that the "entwined" portions remain as such until they are brought onto the supporting film of the EM grid.

The following procedure can be followed. It applies to double-stranded DNA with a GC content of 54-61% (a higher melting temperature must be chosen for higher GC contents).

To be mixed in a plastic tube are (end concentrations)

- 0.045 M sodium phosphate buffer pH 7.0 (made up with double-distilled H_2O),

- 23.5% (v/v) DMSO (warning - solution warms up!),
- 4.5% (v/v) formaldehyde (or formamide - see Sect. 2.3.2.3 c),
- 5-20 μg ml^{-1} DNA,

in an end volume of 0.1-1 ml.

After careful mixing, the specimens are brought to 53°-56° C for 3-8 min; then they are cooled in an ice bath before carrying out one of the various spreading techniques, e.g. the droplet method, which is the quickest and simplest. One should note that a dilution of the melting mixture is necessary in order to have a NA concentration appropriate for spreading. It is advantageous to have a high NA concentration in the melting mixture in order that the end concentrations of DMSO and formaldehyde in the droplet solution be as low as possible.

b) Heteroduplex Techniques

Prerequisite for these methods is the complete separation (melting) of a double-stranded NA into its constituent single-stranded partners. Conditions must then be provided for the creation of hydrogen bonds between homologous segments in a mixture of single stranded NAs from different sources. Nonhomologous sequences, even when they occur within regions of "good fit", will not form double strands and are recognizable as single strands in the EM (see Fig. 2.24 g). Usually it is not possible to achieve a complete separation of the individual partner strands into two separate samples. In such cases a proportion of the double-stranded molecules which are formed after reannealing will represent the original paired strands. This means that searching for and recognizing hybrid molecules can be quite time-consuming. Since NAs have an affinity for plastic surfaces, their adhesion to the walls of reaction vessels (e.g. Eppendorf tubes) must be prevented. This is best done by siliconizing the tubes with either a commercial silicone solution (e.g. Sigmacote SL-2 from Sigma) or with a 1% solution (v/v) of dichlorodimethyl silane in benzene.

1. DNA-DNA-Hybridization

For this type of hybridization DNAs (optimal concentration for each around 0.2 μg ml^{-1}) in a formamide buffer of high ionic strength (40%, v/v, once recrystallized formamide in 0.1 M Tris-HCl pH 8.0 containing 0.01 M EDTA and 0.5 M NaCl) is used. 50 μl of the hybridization mixture is pipetted into an Eppendorf tube, covered with a drop of paraffin oil to prevent evaporation, and heated to 90° C for 5 min (denaturation). This is followed by a second incubation at 37° C (held constant +/−0.1° C) for 2 h (hybridization). Finally the tube is placed in an ice bath. Aliquots of the hybridized mixture can then be processed as above under Section 2.3.2.3.

As an alternative to the above procedure, the following method can be carried out: 30 μl of a denaturating mixture (0.3 μg ml^{-1} DNAs in 0.1 M NaOH and 0.02 M EDTA at pH 8.5) are lightly shaken at room temperature for 10 min. This solution is then neutralized with 3 μl 2 M Tris-HCl pH 7.2. Formamide and NaOH are added to final concentrations of 60% (v/v), incubation is carried out at 32.5° C for 1 h. Once again the reaction is stopped in an ice bath, and the sample is processed as in Section 2.3.2.3.

2. DNA-RNA-Heteroduplex Structures

Because, on the one hand, weakly-buffered formamide solutions tend to decrease in pH and, on the other hand, constant pH conditions (pH 8.0) are necessary for this technique, a buffer system with a high buffering capacity (e.g. 0.1 M Tris or PIPES) is required. DNA (3 μg ml^{-1}) in pH 8.0 buffer is heated to 70° C. When this temperature has been reached, RNA (15–24 μg ml^{-1}; a higher concentration of RNA is necessary in order to achieve a quantitative hybridization with a minimum of DNA-DNA reannealing) is added in a final volume of 100 μl and the mixture maintained at 70° C for a further 10 min to complete the denaturing step. Heteroduplex structures are produced by incubating for at least 30 min at 57°–62° C. Aliquots should be removed at given periods and processed as under Section 2.3.2.3.

3. DNA-RNA-Hybridization Leading to "R-Loop" Structures

In this procedure DNA is present in the form of isolated DNA fragments which have been produced, for example, through the application of restriction endonucleases. The position of certain genes on this DNA is then localized through the application of specific mRNAs. This is achieved by first denaturing the DNA then adding the RNA and finally reannealing the two NA types. Provided RNA is in excess, a significant amount of homologous base pairing will take place, resulting in segments of double-stranded hybrid DNA-RNA. However, under the specific conditions of the incubation there is a good chance that complementary single-stranded DNA may bind to RNA-free segments of the RNA-DNA hybrid. The portion of this DNA which does not bind remains single-stranded and forms a recognizable "R-loop" structure after spreading. Thus genes can be localized by appropriately measuring the length and position of these loops.

The experimental procedure is similar to that described above for the preparation of RNA-DNA-heteroduplex molecules: a high capacity buffer (0.1 M Tris-HCl with 0.01 M EDTA) and a high salt (0.3–0.4 M NaCl) concentration are required. The concentration of formamide is, however, higher (70%, v/v). Denaturing the DNA is attained by heating to 70° C for 10 min; hybridization by adding RNA, covering the mixture (maximum volume 100 μl) with paraffin oil and incubating at a temperature somewhere between 45° and 55° C (note that this temperature is about 6°–8° C lower than that found to be optimal for the production of RNA-DNA heteroduplex structures) are the subsequent steps, followed by processing of the sample as under Section 2.3.2.3.

As a rule the rate of production of RNA-DNA hybrids is primarily dependent upon the RNA concentration. With RNA : DNA proportions of 1 : 2 an incubation period of 30–190 min is to be expected. Smaller RNA concentrations lead to optimal hybridization periods of between 4 and 30 h.

In screening RNA-DNA heteroduplex and R-loop preparations, one often notes that the background is always "contaminated" with RNA. This can lead to difficulties in recognizing the desired hybrid molecule. If this is the case the heteroduplex or R-loop molecules can be isolated from the incubation mixture and concentrated. For this the following procedure can be employed: pack a siliconized glass chromatography column (0.6 × 32–35 cm) with autoclaved Sepharose CL-2B (Pharmacia, separation range 0.7–30 Md spherical particles) to a height of 30 cm. The column is equilibrated with 15 ml of high salt buffer (10 mM Tris-HCl, pH 7.5 – made up with double-distilled H_2O; 1 mM EDTA and 0.8 M NaCl) using a flow rate of 9 ml h^{-1} (hydro-static pressure is equivalent to a column of water 35 cm high). 25–100 μg NA mixture in a volume of 100–150 μl are applied to the column. The RNA-DNA heteroduplex and R-loop structures elute with the void volume, leaving the unhybridized RNA behind. The void volume fraction(s) can be processed directly for electron microscopy as in Section 2.3.2.3.

Literature

Bujard H (1970) Electron microscopy of single-stranded DNA. J Mol Biol 49: 125–137

Chow LT, Broker TR (1978) The spliced structures of adenovirus 2 fiber message and the other late mRNAs. Cell 15: 497–510

Chow LT, Roberts JM, Lewis JB, Broker TR (1977) A map of cytoplasmic RNA transcripts from lytic adenovirus type 2, determined by electron microscopy of RNA:DNA hybrids. Cell 11: 819–836

Davis RW, Simon M, Davidson N (1971) Electron microscope heteroduplex methods for mapping regions of base sequence homology in nucleic acids. Methods Enzymol 21: 413–428

Inman RB (1966) A denaturation map of the λ phage DNA molecule determined by electron microscopy. J Mol Biol 18: 464–476

Küntzel H, Köchel HG, Lazarus CM, Lünsdorf H (1982) Mitochondrial genes in Aspergillus. In: Slonimsky P, Borst P, Attardi G (eds) Mitochondrial genes, Cold Spring Harbor Lab, pp 391–403

Lünsdorf H (1981) Elektronenmikroskopische Transkriptionskartierung des mitochondrialen Genoms aus *Aspergillus nidulans*. Diss, Univ Göttingen

Mayer F, Mazaitis AJ, Pühler A (1975) Electron microscopy of Simian Virus 40 DNA configuration under denaturation conditions. J Virol 15: 585–598

Robash M, Blank D, Fahrner K, Hereford L, Ricciardi R, Roberts B, Ruby S, Woolford J (1979) R-looping and structural gene identification of recombinant DNA. Methods Enzymol 68: 454–469

Wehlmann H, Eichenlaub R (1981) Analysis of transcripts from plasmid mini-F by electron microscopy of R-loops. Plasmid 5: 259–1266

Wellauer PK, Dawid IB (1977) The structural organization of ribosomal DNA in *Drosophila melanogaster*. Cell 10: 193–212

2.3.3 Nucleic Acid-Protein Complexes

2.3.3.1 Specimen Preparation

a) NAs

These must be available in pure form and in a concentration (5–10 µg ml^{-1}) which will allow a dilution without having a negative effect on the spreading method for NA visualization.

b) Proteins

The proteinaceous components (e.g. DNA-dependent RNA polymerases, repressor molecules, restriction endonucleases) must also be available in pure form and in sufficient quantity. They must be present in a buffer free from SDS or other agents which might negatively influence the spreading of the NA. Furthermore, when one is dealing with a nuclease, the binding reaction must be experimentally separable from that of the cleavage. This means that considerable experimentation, particularly with respect to the ionic environment, and especially for divalent cations such as Mg^{2+}, must be undertaken beforehand.

2.3.3.2 Production and Staining of NA-Protein Complexes

In order to achieve the attachment of a protein onto a NA the two are incubated in the appropriate ionic milieu (see above). This can occur either at room temperature or at a higher temperature. Once again, both temperature and concentration have to be optimalized in advance through systematic control experiments. When the incubation has been completed, aliquots are processed via the BAC-method (see Sect. 2.3.2.4); BAC allows for a better recognition of a NA-bound protein than does a cytochrome c procedure. A Pt-Ir-carbon mixture (electron beam evaporation) is used for metal shadowing by the rotation technique.

Experiments in which formaldehyde (end concentration 0.1%, v/v) has been added at the end of the incubation period with the intention of stabilizing the NA-protein interaction tend not to be successful when large numbers of protein molecules become attached to a NA strand. In such cases unspecific coiling occurs, making proper visualization and measurement no longer possible. However, with low numbers of attached protein molecules this rarely occurs.

Literature

Chrysogelos S, Griffith J (1982) *Escherichia coli* single-strand binding protein organizes single-stranded DNA in nucleosome-like units. Proc Natl Acad Sci USA 79: 5803–5807
Gomez B, Lang D (1972) Denaturation map of bacteriohage T7 DNA and direction of DNA transcription. J Mol Biol 70: 239–251

Hamlako BA, Miller OL (1973) Electron microscopy of genetic activity. Annu Rev Biochem 42: 379-396
Hirsch J, Schleif R (1976) High resolution electron microscopic studies of genetic regulation. J Mol Biol 108: 471-490
Johannssen W (1981) Strukturuntersuchungen an der isolierten Restriktionsendonuclease EcoRI und an Komplexen von DNA mit den Restriktionsendnucleasen EcoRI und SAII. Diss, Univ Göttingen
Koller T, Kübler O, Portmann R, Sogo JM (1978) High resolution physical mapping of specific binding sites of *Escherichia coli* RNA polymerase to the DNA of bacteriophage T7. J Mol Biol 120: 121-1131
Zingsheim HP, Geisler N, Mayer F, Weber K (1977) Complexes of *Escherichia coli* lac-repressor with non-operator DNA revealed by electron microscopy: Two repressor molecules can share the same segment of DNA. J Mol Biol 115: 565-570

2.4 Immunoelectron Microscopy (IEM)

2.4.1 Principle Requirements

2.4.1.1 Antigens

The aim of IEM is to identify and localize specific structures in cells, specific components in cell fractions or specific parts of macromolecules. As a rule this spatial allocation is achieved with a resolution which is much better than that which can be obtained by immunofluorescence microscopy. The basic prerequisites for IEM are that the structure in question can be isolated and that it can act as an antigen, i. e. that antibodies can be obtained against it. In addition, the antigen must be stable in a spatial sense, i. e. it should remain where it is both during and after labelling with an antibody. This often means carrying out a chemical fixation before applying the antibody. Ideally, this fixation step should not reduce the ability of the antigens in question to react with the antibodies. As has been shown on numerous occasions, freeze-dried enzyme preparations can be used as antigens for the production of IgG antibodies and the specificity of these antibodies is such that they can, as a rule, be successfully applied to thin sections of fixed material, leading to a highly specific localization of the enzyme.

A further problem to be overcome is that of the access of the antibody to the antigen. This means that the specimen has to be treated in such a way that the antigens are either directly exposed at an external surface, or that the antibodies are able to penetrate into the interior of the structure (cell or organelle). Such treatments should not be allowed to alter significantly the structural integrity of the cell or organelle involved.

2.4.1.2 Antibodies

Both conventional (i.e. polyclonal) and monoclonal antibodies can be used for IBM; usually they are of the IgG type. Antibodies against a particular antigen do not have to represent a pure fraction; it is sufficient when a partially purified IgG fraction contains a significant proportion of the required IgG antibody. The following procedure can be used to separate IgG antibodies (including the desired IgG species) from other immunoglobulins and proteins in blood serum:

Pack a chromatography column with protein A-Sepharose-Cl-4B (Pharmacia) and equilibrate with 0.1 M potassium phosphate pH 7.0. Add 2 ml serum to the column and elute with the same buffer; the IgG antibodies remain bound to protein A in the column. They are then released from the column with either 1 M acetic acid or with glycine buffer (100 mM glycine, 0.4 M NaCl, 0.026% (w/v) NaN_3 at pH 2.8). The IgG fractions can be pooled and stored at −20° C.

Conventional IgG antibodies (see Figs. 2.26 and 2.27e) represent a mixture of antibodies, each specific for a particular epitope on the surface of the antigen in question. Thus, in the case of a protein, it can be expected that a large number of antigenic determinants are exposed. Monoclonal antibodies which are specific for only one particular antigenic determinant allow an even more precise detection and localization of the antigen in question. However, this feature of monoclonal antibodies can be disadvantageous in EM experiments: if the antigenic site is occluded at the moment of application of the antibody no la-

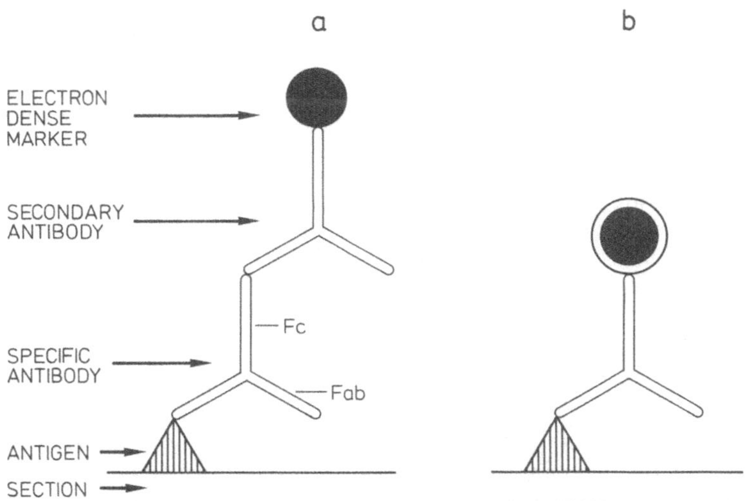

Fig. 2.26a, b. Diagrammatic representation of immunocytochemical labelling methods using colloidal gold as marker. **a** Indirect labelling using an antibody specific for the antigen, a secondary antibody and colloidal gold (without protein A). **b** Direct labelling using a specific antibody complexed with a protein A-coated colloidal gold particle. In this case it is not advisable to use uncoated colloidal gold because this results in a much poorer labelling intensity

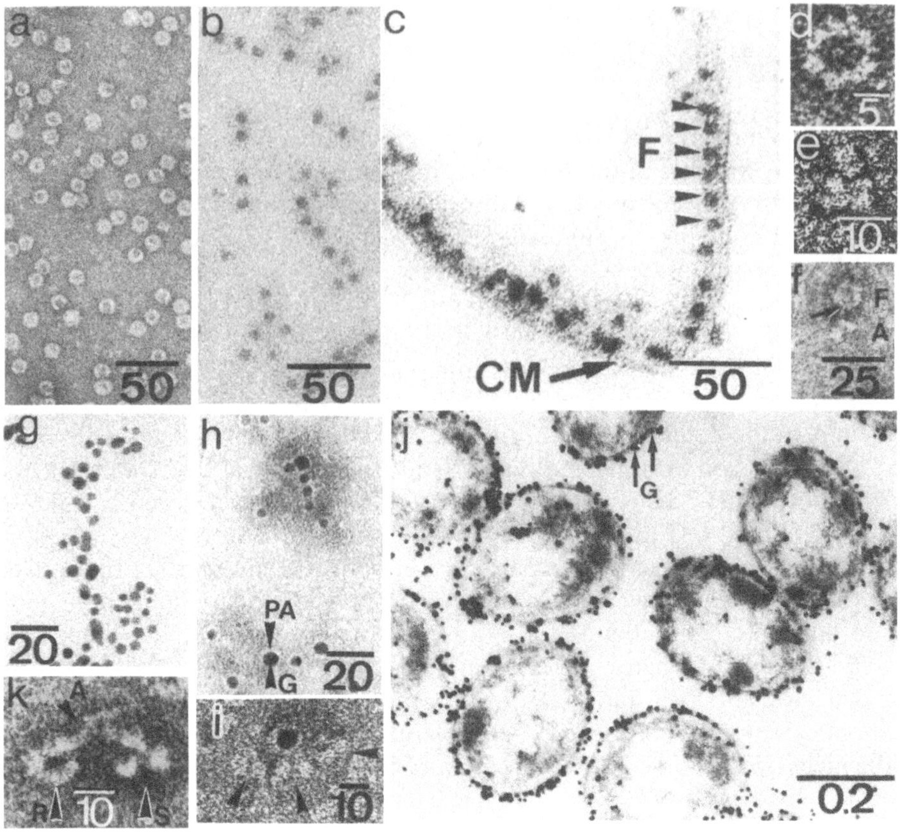

Fig. 2.27. a–d Ferritin molecules. **a, d** Negatively stained with uranyl acetate. The *(dark)* iron core and the *(light)* protein coat (made up of subunits) are clearly visible. **b** Unstained ferritin molecules; the iron core is visible but not the protein coat. **c** Thin section through a segment of the cytoplasmic membrane (cm) of a protoplast of the bacterium *Pseudomonas carboxydovorans* (without cell wall layers); the *arrows* point to ferritin (F) molecules which are lined up at the inner surface of the membrane. The ferritin molecules are attached to goat IgG-antibodies which are specific for rabbit IgG-antibodies which in turn are specific for the antigen (the enzyme carbon monoxide dehydrogenase). Under the conditions of preparation the antibodies are not visible. (Original micrograph M. Rohde).
e IgG-antibody negatively stained with uranyl acetate. The three arms of the Y-shaped molecule can be recognized. The micrograph also hints at the presence of the L- and H-chains comprising each arm. **f** Ferritin(F)-IgG(A) complex negatively stained with uranyl acetate. Complexes of this type become bound to antigen-IgG complexes in antigen localization experiments such as that shown in **c**.
g Unstained colloidal gold. **h** Gold (G) protein A (PA) complexes negatively stained with uranyl acetate. Protein A is visible as an electron-transparent coat around the gold particles. **i** As for **h** but bound IgG antibodies *(arrowheads)*. **j** Thin section through cells of the bacterium *Pseudomonas carboxydovorans* showing the immunocytochemical localization of the enzyme carbon monoxide dehydrogenase with the help of protein A-gold. Poststaining with uranyl acetate and lead citrate.
k Localization of the large subunits in the enzyme citrate lyase (see also Figs. 2.20 and 4.1). *A* IgG-antibody specific for the large subunits; *R* "ring" form of the enzyme complex; *S* "star" form. Two arms of the antibodies are seen bound to major elements of the R and S structures, indicating the presence here of the large subunits. Negative staining with uranyl acetate (Zimmermann et al. 1982). Magnification bar in **j** in μm; otherwise in nm

belling can occur. In contrast, when conventional, polyclonal, antibodies are applied there is always a good chance that at least some of the antigenic determinants will remain exposed and can be labelled.

2.4.2 Labelling of Antigens in Cells and Cell Fractions

2.4.2.1 Ferritin-Labelled Antibodies

Antigens can be localized by applying antibodies "tagged" with an electron-dense marker. Ferritin, because of its iron core, is such a substance (see Fig. 2.27 a, b, d). It can be attached directly to IgG antibodies although this can lead to a loss of antibody reactivity or to the production of antibody-ferritin aggregates or even ferritin-ferritin complexes. A further problem with ferritin-conjugated antibodies is that the ferritin can still bind unspecifically to a variety of cellular constituents, thereby casting a shadow of doubt on localization experiments.

Often the antibody is labelled indirectly with ferritin via a second IgG-antibody. This antibody has as its antigen the first antibody, i.e. it is an anti-antibody. Since the ferritin is coupled to the anti-antibody it cannot interfere with the initial interaction between antigen and primary antibody. This leads to a higher specificity of localization. Despite the presence of this second antibody, the precision of this method is hardly affected (see Fig. 2.27 c).

Another form of indirect immunolabelling with ferritin involves the protein avidin (streptavidin) rather than a second antibody to a primary IgG. Avidin is a protein present in egg white which binds tenaciously to biotin (four binding sites per molecule). Thus specific antibodies which are linked to biotin can be visualized in a two-stage method. First of all avidin is given to the biotinylated antibody-antigen complex, and then biotinylated ferritin is added. This binds to one of the three vacant binding sites of the avidin and leads to an increased visibility of the antigen-antibody complex. Theoretically two molecules of biotin bound to one primary antibody can lead to the binding of five ferritin molecules via two avidins.

Immunoferritin Labelling of Organelle Suspensions

250 μl of a particle or organelle suspension are mixed with either 250 μl potassium phosphate buffer (0.1 M, pH 7.0) or 0.6 M sucrose containing 10 mM $MgCl_2$ in an Eppendorf tube. An aliquot of the IgG fraction is added (0.5 mg per 500 μl buffered organelle suspension) and the mixture incubated at 37° C for 2 h with occasional gentle shaking. The mixture is then centrifuged for 5 min at 18000 *g* (at least) and the pellet washed twice by resuspending and centrifuging in the above buffer. The washed pellet is resuspended in 250 μl buffer (or sucrose solution) and reincubated for 2 h at 37° with an anti-IgG-ferritin conju-

gate (commercially available). After centrifuging and washing the pellet can be fixed, dehydrated and embedded. This procedure represents a pre-embedding labelling.

Literature

Acker G, Knapp W, Wartenberg K, Mayer H (1981) Localization of enterobacterial common antigen in *Yersinia enterocolitica* by the immunoferritin technique. J Bacteriol 147: 602-611
Bayer EA, Wilcheck M, Skutelsky E (1976) Affinity cytochemistry: the localization of lectin and antibody receptors on erythrocytes via avidin biotin complex. FEBS Lett 68: 240-244
Breese SS, Hsu KC (1971) Techniques of ferritin-tagged antibodies. In: Maramorosch K, Koprowski H (eds) Methods in virology, vol V. Academic Press, London New York, pp 399-422
Driel D van, Wicken AJ (1973) Cellular location of the lipoteichoic acids of *L. fermenti* NCTC 6991 and *L. casei* NCTC 6375. J Ultrastruct Res 43: 483-497
Johnstone A, Thorpe R (1982) Immunochemistry in practice. Blackwell, Oxford pp 1-298
Perrot-Rechenmann C, Gadal P (1986) Enzyme immunocytochemistry. In: Wang TL (ed) Immunology in plant science. Cambridge Univ Press, Soc Exp Biol Sem Ser 29: 59-88

2.4.2.2 Immunolabelling with Protein A-Gold

This technique is especially useful for the localization of antigens in thin sections. Since the labelled antibody is added to a section this procedure represents a post-embedding immuno-localization method. This method puts to use the fact that protein A (from *Staphylococcus aureus*) binds specifically to the Fc portion of IgG-antibodies (see Figs. 2.26 and 2.27 i). When protein A is combined with a colloidal gold particle the site of the antibody-antigen complex can be excellently visualized in the EM.

Preparation of Colloidal Gold Particles

Two methods are currently in use. In the first, 2.5 ml of a 0.6% (v/v) tetrachloroauric acid solution are added to 60 ml filtered, double-distilled water in a siliconized (see Sect. 2.3.2.5 b) 100-ml Erlenmeyer flask. 1 ml of a phosphorus solution (prepared by adding 8 vol diethyl ether to 2 vol. of a white phosphorus-saturated ether solution – note: dissolving the latter takes 2-3 days) is then slowly pipetted into the stirred gold chloride solution. The mixture is shaken strongly and allowed to stand for 15 min at room temperature, before boiling for 5 min in a reflux condensor. The resulting gold sol has a typical wine-red colour and is brought to pH 6.9 with 0.2 M K_2CO_3 (use pH paper). Gold particles of a uniform size (see Fig. 2.26 g) can be obtained from this sol by centrifugation. Thus the solution is centrifuged at 100,000 *g* for 1 h (Beckman Airfuge, A-100 Rotor); 3-8 nm particles are then found to be present in the lowermost quarter of the centrifuge tube. They can be stored at 4° C for several weeks.

A simpler method is as follows: 1 ml 1% (w/v) tetrachloroauric acid, 4 ml 1% (w/v) sodium citrate, 0.5 ml 1% tannic acid (Important! For this method one must use low molecular weight gallolyl glucose available from Mallinckrodt, St. Louis, USA, Code Nr. 8835), 0.5 ml 25 mM K_2CO_3 and 94 ml triple-distilled water are mixed together in a siliconized 250 ml Erlenmeyer flask and stirred on a hot plate at 60° C. Within seconds the colour changes to wine-red, indicating the formation of a gold sol. The mixture is then boiled for a few minutes and allowed to cool to room temperature. Colloidal gold particles of larger or smaller diameter can easily be obtained by altering the volume of tannic acid in the reducing mixture, e.g. 2ml 1% tannic acid: 4 nm; 0.125 ml: 8 nm; 0.03 ml: 11 nm.

Coupling Colloidal Gold to Protein A

Before attempting to prepare a protein A-gold complex it is necessary to determine in advance the minimum amount of protein A which can stabilize a gold sol in the presence of a strong electrolyte. This can be done by mixing together 100 µl of a protein A solution (Pharmacia; a concentration series is prepared from a stock solution of 1 mg ml^{-1} protein A). The mixture is allowed to stand at room temperature for 5 min before adding 0.5 ml of a 10% (w/v) NaCl solution and leaving to stand at room temperature for a further 10 min. If insufficient protein A is present, clumping of gold particles occurs and a blue colour will develop. If a light red colour develops (absorbing at 580 nm) the colloidal gold particles can be judged to be stable. The concentration of protein A thus determined (plus a safety margin of 20%) can be used to prepare the protein A-gold complex for immunogold labelling. Of this solution (aqueous) 0.2 ml is mixed with 10 ml gold sol and left to stand for 10 min at room temperature. 1 ml of a 1% (w/v) polyethylene glycol (PEG) solution (MW 20,000) is then added and the mixture centrifuged at 100,000 *g* for 1 h. The protein A-coated colloidal gold particles are removed from the lowermost quarter of the centrifuge tube and taken up in PBS (phosphate buffered saline: 50 mM potassium phosphate buffer pH 7.0 containing 0.9%, w/v, NaCl) containing 0.2 mg ml^{-1} PEG. This solution is centrifuged again and taken up a second time in PBS-PEG solution. The protein A-gold solution finally obtained (see Fig. 2.27 h) can be stored at + 4° C for several months.

Immunogold Labelling of Plastic Sections

a) Pretreatment of the Sections

Gold-coloured sections are collected on PBS (see above) instead of water, and are not stretched with chloroform vapour. They are picked up on nickel grids without a support film. In order to reveal a sufficient number of antigenic sites, the sections are then etched by floating (sections facing downwards) on a drop of 10% (v/v) H_2O_2 or 10% (v/v) sodium methylate for 15 min. After washing well with PBS, the sections are placed on a drop of saturated sodium metaperiodate which counteracts the masking effect of OsO_4 on antigenic sites. This is followed by washing again in PBS. Treatment with H_2O_2 can be omitted; treatment with sodium metaperiodate should only be used when OsO_4 has been applied beforehand. In all of these and in the subsequent steps it is important that the sections are not allowed to dry out.

b) Incubation with the Antibodies

The sections are laid onto a drop of IgG antibody solution (a series of concentrations, diluted with PBS, should be set up) on a piece of parafilm and left overnight at 4° C in a covered Petri dish. The sections are then thoroughly washed with PBS (at least two 50-ml beakers of PBS should be used) followed by three 10-min transfers on drops of PBS.

c) Incubation with Protein A-Gold

Drops (50 μl) of protein A-gold solution are placed onto a piece of parafilm in a Petri dish. The sections from (b) are placed onto the drops and incubated at room temperature for 3 h. Finally the sections are thoroughly washed, first in PBS, then in filtered double-distilled water. For poststaining the sections see Section 2.2.6.3

Immunogold Labelling of Cryosections

In this method it is important that, although the sections should remain wet, the reverse side of the grids must always be kept dry. This is normally achieved, as above, by floating the grids, with the sections downwards, on drops of the respective solutions. In addition, 1% (w/v) BSA is usually added to all solutions in order to prevent unspecific background labelling.

The moisture chamber (see Sect. 2.2.5), previously maintained at ice temperatures, is brought to room temperature. After several minutes the gelatine begins to melt. Thus the grids, with sections facing downwards, now float at the surface of the liquefied gelatine. Each grid is then washed three times by transferring to drops of PBS (here containing additionally 0.02%, w/v, KCl). After being washed, the grids are placed onto small drops (5 μl) of the IgG antibody in PBS (concentration between 10 and 200 $\mu g\ ml^{-1}$) and incubated (drops on parafilm, closed Petri dish) for 30 min at room temperature. The grids are then washed four times by transferring to drops of PBS each for 1 min. Finally, each grid is placed onto a drop of protein A-gold solution and incubated for 30 min at room temperature before being washed thoroughly with PBS (4 drops each 5 min).

Literature

Bendayan M (1983) Ultrastructural localization of actin in muscle, epithelial and secretory cells by applying the protein A-gold immunocytochemical technique. Histochem J 15: 39-158

Bendayan M, Zollinger M (1983) Ultrastructural localization of antigenic sites on osmium-fixed tissues applying the protein A-gold technique. J Histochem Cytochem 31: 101-1109

Bendayan M, Roth J, Perrelet A, Orci L (1980) Quantitative immunocytochemical localization of pancreatic secretory proteins in subcellular compartments of the rat acinar cell. J Histochem Cytochem 28: 149-160

Griffiths G, Simons K, Warren G, Tokuyasu KT (1983) Immunoelectron microscopy using thin frozen sections: application to studies on the intracellular transport of semliki forest virus spike glycoproteins. Methods Enzymol 96: 466-483

Hjelm H, Hjelm K, Sjöquist J (1972) Protein A from *Staphylococcus aureus*. Its isolation by affinity chromatography and its use as an immunosorbent for isolation of immunoglobulin. FEBS Lett 28: 73-76

Horisberger M, Rosset J (1977) Colloidal gold, a useful marker for transmission and scanning electron microscopy. J Histochem Cytochem 25: 295-305

Rohde M (1983) Immunoelektronenmikroskopische Untersuchung der in situ Lokalisierung von Kohlenmonoxid-Oxidase aus *Pseudomonas carboxydovorans* mit Hilfe der Immunferritin- und der Protein A-gold Technik. Diss, Univ Göttingen

Roth J (1982) The protein A-gold (pAg) technique. Qualitative and quantitative approach for antigenic localization on thin sections. In: Bullock GR, Petrusz P (eds) Techniques in immunochemistry, vol 1.Academic Press, London New York, pp 103-135

Roth J, Bendayan M, Orci L (1978) Ultrastructural localization of intracellular antigens by the use of protein A-gold complex. J Histochem Cytochem 26: 1074-1081

Slot JW, Geuze JH (1984) Gold markers for single and double immunolabelling of ultrathin cryosections. In: Polak JM, Varndell IM (eds) Immunolabelling for electron microscopy. Elsevier, Amsterdam, pp 129-142.

2.4.3 Localization of Protein Subunits with Specific IgG Antibodies

The isolated, purified protein is dissociated into its subunits which are then separated and collected by preparative gel electrophoresis. Antibodies are prepared against the various polypeptides. (Interestingly the presence of SDS in the protein sample does not reduce the specificity of the antibodies). Preimmune serum is also prepared from blood taken from the animals which are later to be injected. In order to check the specificity of the IgG-fraction, double diffusion tests or Western Blotting are carried out with the respective antigen; no reaction should be obtained with the antibodies of the preimmune serum, nor should there be a cross reaction with other polypeptide antigens.

2.4.3.1 Preparation and Visualization of the Protein-Antibody Complex

1 mg of the original protein is mixed with 0.5 mg IgG protein in the presence of 0.1 M potassium phosphate buffer pH 7.0 (end volume 300-500 μl; the presence of divalent cations may also be necessary) and incubated at 30° C for 30 min. The mixture is then put onto a Biogel-A-1.5 m column (0.9 × 58 cm) pre-

viously equilibrated with phosphate buffer. Fractions of 0.75 ml are collected and processed for electron microscopy by negative staining (see Fig. 2.27 k).

An alternative way of separating antibody-protein complexes from the individual components is to centrifuge the mixture in a 5-30% (w/v) linear sucrose gradient to isopycnic conditions (14 h; 4° C; 160,000 *g*). Aliquots from 0.5 ml fractions can then be negatively stained.

One hopes that the contact position (the Fab region) on the Y-shaped IgG antibody on the respective antigenic polypeptide subunits is recognizable (the size of an IgG antibody in negative contrast is around 12-13 mm).

Literature

Bowien B, Mayer F (1978) Further studies on the quaternary structure of D-ribulose-1,5-bisphosphate carboxylase from *Alcaligenes eutrophus*. Eur J Biochem 88: 97-107

Tischendorf GW, Zeichhardt H, Stöffler G (1975) Architecture of the *Escherichia coli* ribosome as determined by immune electron microscopy. Proc Natl Acad Sci USA 72: 4820-4824

Wabl MR (1974) Electron microscopic localization of two proteins on the surface of the 50 S ribosomal subunit of *Escherichia coli* using specific antibody markers. J Mol Biol 84: 241-247

Zimmermann Th, Giffhorn F, Schramm HJ, Mayer F (1982) Analysis of structure-function relationships in citrate lyase isolated from *Rhodopseudomonas gelatinosa* as revealed by cross-linking and immunoelectron microscopy. Eur J Biochem 126: 49-56

2.5 Autoradiography

2.5.1 General Background

There are two major prerequisites for the performance of autoradiographic investigations:

1. substances labelled with a suitable radioactive isotope must be available;
2. once taken up by the cell, these substances must emit a form of radiation which is easily detectable.

Autoradiography, therefore, allows one to determine the movement and fate of a variety of substances, especially metabolites, in the cell.

2.5.1.1 Physical Basis

Radioactive isotopes are atoms in an unstable state. Through disintegration of the nucleus they try to reach an energetically stable state and thereby emit energy in the form of radiation. Of the three types of radiation (α, β or γ) which can be emitted, only the β form which consists of negatively charged particles (electron radiation) is of interest for autoradiography. Not all β-radiating isotopes can be employed in autoradiographic investigations. There are those which emit electrons with so much energy that the radiation is still detectable several meters distant from the source.

Conversely, there are those whose emitted electrons possess so little energy that, even within a few micrometers from the source, they are no longer detectable. Clearly when the position of an isotope is to be determined as a result of its radioactive properties, it is better to choose an isotope whose zone of detection is not too large. This is the basis for resolution in autoradiography: the smaller the energy of a β-radiating isotope is, the more precisely can its position be determined. To illuminate this point we can compare two isotopes regularly in use in autoradiography: tritium (^3H) and radioactive carbon ^{14}C). The maximum β-energy of the former is 0.0186 MeV (megaelectron volt) which is such that radiation can no longer be detected at a distance greater than 2–3 μm from the isotope. ^{14}C has a β-energy of 0.156 MeV (Table 2.11) which means that electrons can still be detected at a distance of more than 20 μm from the isotope. Thus when one is able to detect a tritiated substance one can be sure that the true position of the isotope is within 2–13 μm; with ^{14}C-labelled substances the resolution is up to ten times poorer. However, as we shall see, other factors are important in determining resolution in autoradiography.

Another property of radioactive isotopes, which must be kept in mind when considering their applicability for autoradiography, is their half-life. The speed with which the nucleus disintegrates is not the same for all isotopes. This is reflected in their so-called half-lives which are defined as the time in which the radiation of an isotope has diminished to one half of its original amount (see Table 2.11 for values). Since the preparation of sections for electron microscope autoradiography takes several days it can hardly be recommended to use an isotope whose half-life is also a matter of days. This becomes even more important when one takes into consideration the fact that the amount of isotope present in a thin section is often so small that one already has difficulties in detecting the emitted radiation. Usually one employs isotopes whose half-lives are such that there is hardly any change in the level of radiation emitted over a 6-month period. This is the case for ^3H (half-life 12 years) and ^{14}C (half-life 5730 years).

β-energy and half-life describe the properties of an isotope in a qualitative way. The quantity of an isotope present is expressed in terms of the number of nuclear disintegrations which occur over a given period of time. Two units are

Table 2.11. Half-life times and β-energy values of isotopes used in autoradiography

Isotope	Half-life	Max. β-energy MeV
^3H	12.35 years	0.0186
^{14}C	5730 years	0.156
^{35}S	87.4 days	0.167
^{32}P	14.3 days	1.709

Table 2.12. Units of radioactivity

1 Becquerel (Bq)	= 1 disintegration s^{-1}	
1 Curie (Ci)	= 3.7×10^{10} Becquerel	= 37 Gigabequerel (GBp)
1 Millicurie (mCi)	= 3.7×10^{7} Becquerel	= 37 Megabecquerel (MBq)
1 Microcurie (μCi)	= 3.7×10^{4} Becquerel	= 37 Kilobecquerel (kBq)
1 Gigabecquerel (GBq)	= 27.027 mCi	
1 Megabecquerel (MBq)	= 27.027 Ci	
1 Kilobecquerel (kBq)	= 27.027 nCi = 0.027027 Ci	

in use which describe the "radioactivity" of an isotope. The official (SI) unit is the Bequerel (Bq) which describes the amount of isotope giving rise to one disintegration per second. The older unit, which is still used by many workers, is the Curie, which corresponds to 3.7×10^{10} disintegrations per second. Depending on the amount of isotope, these units may be pre-fixed with a decimal term indicating larger or smaller values (see Table 2.12).

2.5.1.2 Chemical Basis

Radioactive isotopes are presented to cells much less frequently than molecules in which one or more atoms are radioactive. A knowledge of the chemistry of such compounds is therefore just as important in autoradiography as an appreciation of the physical parameters of an isotope. This is particularly so when one considers that radioactive compounds are often metabolized once they enter the cell. Indeed, autoradiography is a method of choice for establishing the subcellular sites of particular steps in metabolic pathways. Thus one can study DNA synthesis (see Fig. 2.28 a) and protein synthesis by incubating cells with radioactive thymidine and amino acids, respectively. However, the quality of the results obtained is invariably dependent upon the specificity of the precursor involved.

a) Specificity of the Radioactive Precursor

This can best be illustrated by experiments to localize the site of synthesis of carbohydrate side chains in glycoproteins. Two precursors are available for this purpose: glucosamine and fucose. Whereas the latter is more or less specifically incorporated as a terminal sugar in the oligosaccharide side chains of glycopro-

Fig. 2.28. a Demonstration of DNA synthesis in the nucleus (N) of a mouse cell grown in vitro. Labelling with 20 μCi ^{3}H-thymidine (Sp. Act. 5 Ci $mmol^{-1}$) culture medium for 1 h. Exposition time = 55 days; compact development. **b** Demonstration of intracellular transport in a tubule cell of a mouse kidney. Glycoproteins labelled with 50 μCi ^{3}H-fucose (Sp. Act. 15 Ci $mmol^{-1}$) per mg body weight for 8 h. The products which were initially synthezised in the Golgi apparatus have, after 8 h, reached the cell surface (indicated by *arrowheads*). *B* basal membrane of the tubule; *N* nucleus. Exposition time = 70 days; compact development. Magnification bar in μm

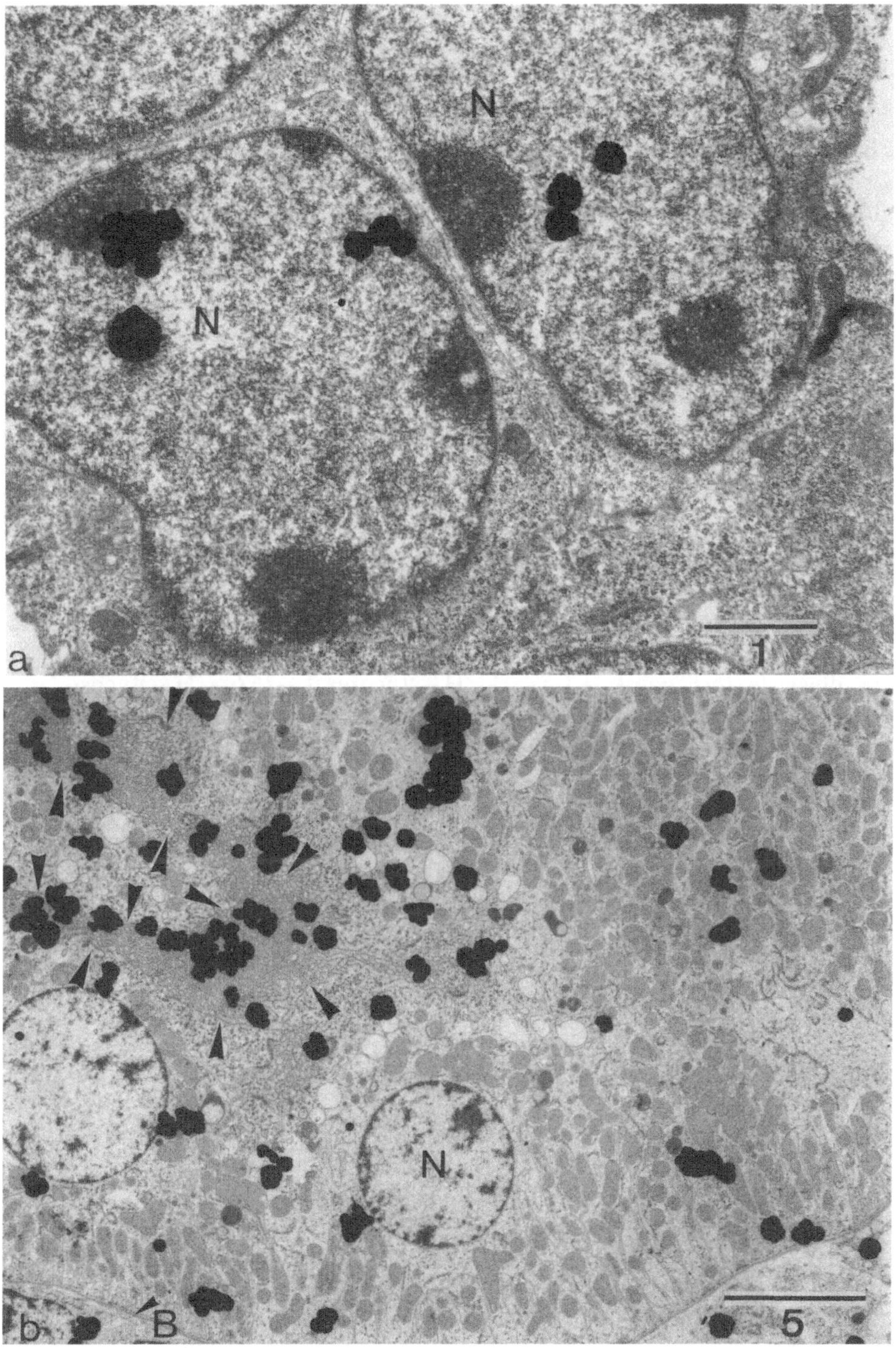

Fig. 2.28 a, b

teins, glucosamine is also incorporated into glucosaminoglycans. Thus fucose would be in this case the precursor of choice.

b) Specific Activity of the Precursor

Because of their better resolution, tritiated compounds are used most frequently in autoradiographic investigations. Such compounds are usually synthesized through exchanging hydrogen atoms with tritium. Thus the greater the exchange the greater the radioactivity in the molecule. The relative amount of radioactivity in a compound is expressed in terms of Bq or Ci per g (i. e. in terms of mass) or per mole or m mole (i. e. in terms of molecular weight). This, however, does not mean that each molecule possesses the same amount of isotope; instead the compounds which one purchases are usually mixtures of strongly and weakly labelled molecules, together with a proportion of unlabelled molecules. The specific activity which is given on the package is thus an average of this mixture.

c) Specificity of Labelling

Radioactive molecules are usually bought as universally or specifically labelled substances. In the former case all the atoms of a particular type, e.g. carbon, have been replaced with the isotope ^{14}C. In the other case only specific atoms have been exchanged. Selecting the correct specifically labelled molecule is often most important. Again using glycoprotein biosynthesis (see Fig. 2.28 b) to illustrate this point: L−(−1−^{3}H−)fucose, L−(5,6−^{3}H−)fucose and L−(6−^{3}H−)fucose can all be used for an autoradiographic investigation of glycoprotein biosynthesis, but not L-(1-^{3}H-)fucose, since the attachment of fucose to the oligosaccharide sidechain occurs via the C_1 atom and this results in the release of the tritium from the molecule.

Literature

Catalogues of the various companies selling radionuclides, e.g. Amersham, New England Nuclear

Rogers AW (1979) Techniques of autoradiography. Elsevier, Amsterdam, pp 1-430

2.5.2 Choice and Dosis of Radioactive Compounds

2.5.2.1 Choosing a Radioactive Precursor

In addition to considerations of the biochemical suitability of a precursor together with its appropriate labelling characteristics there are other factors which have to be taken into account when selecting the best precursor for an autoradiographic investigation. As already mentioned, tritiated compounds provide for the best resolution of a radioactive source but there are a number of reasons why, in some cases, ^{3}H-labelled compounds are not to be preferred. First of all, and most simply, the desired tritiated labelling combination may not be available. Secondly, the tritiated compound may not be available in the specific activity which is necessary for adequate detection in tissues by autoradiographic means. Thirdly, tritiated compounds have the tendency, in aqueous solution, to exchange back their ^{3}H atoms for normal H atoms. This can be reduced greatly by the presence of ethanol, indeed some tritiated compounds are only stable in concentrated ethanolic solutions. For example, the tritiated form of the carbohydrate precursor uridine diphosphoglucose (UDP-glucose) is only available in a 50% ethanolic solution. Because of the effects of ethanol such solutions cannot usually be administered directly to living tissues.

As a rule ^{14}C-labelled molecules make for the best alternatives to tritiated compounds, since these are normally available in pure water and possess enough β-energy for ease of detection. However, other isotopes are occasionally met, e.g. ^{35}S, which is often used in studies on protein synthesis. Although ^{35}S has a relatively short half-life (87 days) the specific activity of ^{35}S-methionine, for example, is at 600 Ci $mmol^{-1}$ almost ten times higher than ^{3}H methionine (max. Sp. Act. 70-85 Ci $mmol^{-1}$) and therefore allows detection at much lower amounts than is possible with the tritiated form.

Finally, there are two points well worth considering in the choice of a suitable radioactive precursor. The first involves possible negative effects of the precursor on cell metabolism. This can best be illustrated with thymidine as a precursor for DNA synthesis. Cells can be arrested in the S phase through a blockage of DNA synthesis when thymidine is given to cells at too high a concentration. The only way to avoid this is to present the cells with low concentrations of thymidine of a high specific activity. The second point concerns the nature of the substance/molecule into which the isotope of the precursor is finally incorporated. If the isotope is in soluble form it is likely that much of it will be lost by leaching during fixation and embedding. This is also a possibility for substances which are only loosely bound, e.g. radioactive ligands bound to receptors at the cell surface or within the endomembrane system. Clearly only those radioactive molecules which are integrated into stable structures in the cell stand a good chance of being detected at the end of the preparation procedure for electron microscopy.

Literature

Alberts B, Bray D, Lewis J, Raft M, Roberts K, Watson ID (1983) Molecular biology of the cell. Garland, New York, pp 1-1146
Lehninger AL (1975) Biochemistry. Worth, New York, pp 1-1104
Michael G (1976) Biochemical pathways. Boehringer, Mannheim

2.5.2.2 ***Dosage***

Although there are no tables available from which one can read off the necessary amounts of radioactivity required for a successful autoradiographic investigation in a particular cell or tissue, there are some useful guidelines.

Particularly when working with animals, it is important to take into account the fact that some biochemical events are strictly localized in one particular organ, whereas others may be ubiquitous in all of the cells. In the latter case a much higher dosis of radioactive precursor (e.g. amino acids when studying protein biosynthesis) must be given to the animal in order to produce satisfactory autoradiographic results than, for example, when studying the synthesis of thyroglobulin. If, in the latter case, a suitable precursor is given to the animal, it will selectively accumulate in the cells of the thyroid gland. This means that, here, smaller amounts of radioactivity will suffice.

Whereas in LM-autoradiography the β-energy of the isotope in question has to be taken into consideration (see Sect. 2.5.1.1), this is not as important in determining the dosage of a radioactive substance for EM-autoradiography. Instead of using ten times less ^{14}C than ^{3}H, as in LM-autoradiography, one normally sets the two isotopes in a 1:2 relationship respectively for EM-autoradiography. Of course in so doing one must not forget to take into account the specific activity of the isotopes involved, e.g. 2 mCi ^{3}H-glucose (Sp. Act. 1 mCi mmol^{-1}) corresponds to 0.2 mmol glucose.

As a rule LM sections are about 20 times the thickness of a good TEM section (thickness 50 nm). This means that sections for LM-autoradiography contain roughly 20 times as much radioactivity as sections for TEM. Hence for the same effectivity in terms of detection and resolution 20 times higher amounts of radioactivity should be presented to cells for TEM-autoradiography than is the case for LM-autoradiography. Amounts are best expressed in terms of fresh weight (μCi g^{-1}) or volume of culture medium (μCi ml^{-1}) for cells in suspension. Extrapolations from in vivo to in vitro conditions and vice versa can normally be done on the basis that 1 g fresh weight and 1 ml culture are equivalent to one another.

An actual example will now be given to illustrate the foregoing. In studies of DNA synthesis thymidine is the usual precursor. For SM-autoradiography 1–2 μCi g^{-1} fresh weight (1–2 μCi ml^{-1} suspension culture) of ^{3}H-thymidine will produce satisfactory results. For EM-autoradiography 20–40 μCi g^{-1} or ml^{-1} are necessary. If in the latter case ^{14}C-thymidine is preferred, 10 μCi g^{-1} will then be required.

2.5.3 Working with Isotopes-Radiation Protection

A special permit for working with isotopes is required in almost every country. The type of isotope normally determines the extent of the safety measures necessary. Autoradiography belongs to the less hazardous of experiments involving isotopes. In the case of low energy (up to 1.0 MeV) β-isotopes in aqueous solutions the danger of direct radiation is negligible (β-radiation from ^{3}H and ^{14}C can only reach 5–6 μm and 0.2–0.3 mm, respectively). However, care must be taken to avoid direct contact with the skin or ingestion of such isotopes. Thus plastic gloves should be worn and pipetting should not be done with the mouth. For β-isotopes with higher energy, which find occasional use in autoradiography, e.g. ^{32}P (1.709 MeV) additional safety precaution are necessary, e.g. the experiment should be performed behind a perspex shield and protective glasses should be worn when working with isotopes such as ^{131}I which, although being a weak β-emitter (0.25–0.8 MeV), also sends out γ radiation.

2.5.4 Preparation of Radio-Labelled Cells/Tissues for Electron Microscopy

As already mentioned under Section 2.5.2.1, it is important to know whether an isotopically cellular molecule is firmly or weakly bound to a cellular structure or whether it is present, unbound, in a soluble form in the cell. If this is not already known it can easily be determined by homogenizing a portion of the labelled tissue and centrifuging at 100,000 *g* for 30 min to produce a pellet (cell organelles) and a supernatant (cytosol). The radioactivity in both pellet and supernatant is then determined by scintillation spectrometry. Should more than 50% of the radioactivity lie in the pellet, one can assume that a major position of the isotope is present in a "bound" form. In this case, despite some losses during fixation and dehydration, the normal procedures for fixing and dehydration described in Section 2.1 can be employed. When, however, a large proportion of the radioactivity is found in the supernatant, there is little chance that a sufficient amount will be present in the organelles after embedding. This means that other methods of specimen preparation are necessary.

In order to avoid, on the one hand, the diffusible loss of soluble radioactive compounds and, on the other, to prevent their dislocation within the cell, cells or tissues are best freeze-fixed as described in the chapter on freezeetching (see Sect. 2.6). The frozen material can then be processed for cryo-ultramicrotomy (see Sect. 2.2.5) or freeze-dried. The latter method is used more often in the preparation of tissues containing soluble radioactive compounds for autoradiography. In this method the frozen specimen is transferred onto a precooled object stage at $-85°$ C in a freeze-drying apparatus whose vacuum is better than 10^{-6} torr. When dry, the specimen is placed in a metal meshwork holder which is suspended over a beaker containing buffered OsO_4 (2%, w/v). This is left to stand in an exsiccator for 24 h at room temperature. After the osmium fume-fixation the specimen can be embedded directly in an epoxy resin. Since freeze-dried specimens tend to be hygroscopic, the greatest danger in this procedure is the absorption of water vapour from the air during the transfer steps. This problem is overcome in many modern freeze-drying machines which are equipped to carry out all three steps (drying, fixing and embedding) without interrupting the vacuum.

Literature

Stumpf WE (1976) Techniques for the autoradiography of diffusible compounds. In: Prescott D (ed) Methods in cell biology, vol XIII, chap 9. Academic Press, London New York, pp 171–193

Terracio L, Schwabe KG (1981) Freezing and drying of biological tissues for electron microscopy. J Histochem Cytochem 29: 1021-1028
Wilske KR, Ross R (1965) Autoradiographic localization of lipid- and water-soluble compounds: a new approach. J Histochem Cytochem 13: 38.43

2.5.5 Photographic Emulsions and Autoradiography

2.5.5.1 Apparatus Required

Not only are some special pieces of apparatus necessary for performing autoradiography; a dark room is also an absolute necessity. One must be certain that the development of silver grains in the photographic emulsion is not a consequence of diffuse light. Whenever possible, this dark room should be used exclusively for autoradiographic purposes. To avoid dust accumulation it should contain as little furniture as possible and ideally should have washable walls and ceilings. Apart from a stool and a cupboard for chemicals, the furniture should be restricted to a photographic bench with sink, a table for coating the EM grids with photoemulsion, and a refrigerator (with low temperature freezer) for exposing the emulsion-covered sections.

The room should be fitted out with hot and cold running water and have a sufficient number of electrical sockets and lamps. When using Ilford L4-emulsion, a normal red filter should be fitted on each lamp, but one should be aware that this may have to be changed for different emulsions (e.g. for Kodak NTE a Wratten Series # 2 Filter is necessary). Since photographic emulsions are sensitive to temperature changes, it is advisable to have the room climatized at 21°-22° C and with 50% humidity. If this is not possible one should at least avoid having the room exposed to direct sunlight, i.e. choose a room in the basement or interior of the building.

The following small pieces of apparatus are also required: a small water bath (2 l) with temperature regulator (35°-50° C); and an assortment of small (20, 50, 100 ml) beakers and glass rods. Additional special equipment will be given below.

2.5.5.2 Choice of Emulsion; Consequences for Resolution

Photographic emulsions consist of silver bromide grains embedded in a gelatine layer. Electrons which are emitted from a radioactive source cause a reduction of the silver bromide, leading to a deposition of silver grains. Emulsions differ in terms of silver grain size and density. These factors also control the sensitivity of an emulsion. Although for purposes of resolution in autoradiography extremely sensitive emulsions are to be preferred, paradoxically, high resolution can only be achieved at the cost of sensitivity. This can be explained in the following manner.

Let us assume that the tritiated compound is incorporated in a cell into a 1 nm diameter object and that this object lies at the surface of the section. Radi-

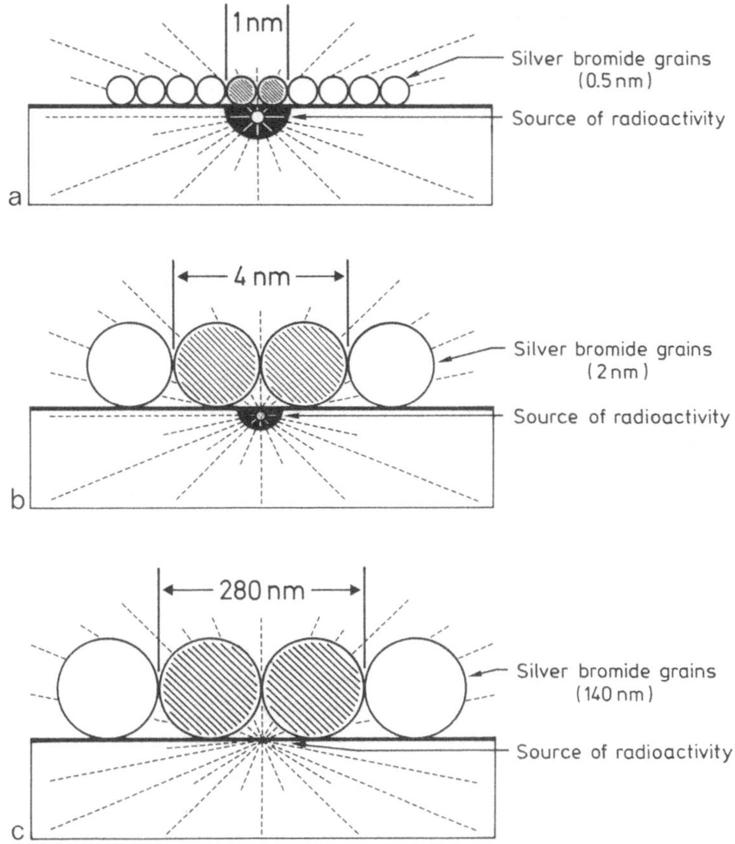

Fig. 2.29. a–c Relationship between the size of a silver bromide crystal and resolution in EM-autoradiography. Two adjacent silver bromide crystals will most likely become exposed when they lie above the radioactive source. If the silver bromide crystals have a diameter of 0.5 nm **(a)** the resolution will then be 1 nm; if the crystals have a diameter of 2 nm the resolution becomes 4 nm; if the crystals have a diameter of 140 nm (Ilford L4 emulsion) the resolution will become 280 nm. Only when a crystal lies directly above the source can the resolution with Ilford L4 emulsion be reduced to 140 nm

ation is emitted equally in all directions from this source. Ideally, the source will be optimally localized when two 0.5 nm silver grains, which lie directly over the source, are developed. If the silver grains have a diameter of 2 nm the resolution of the radioactive source will become 4 nm (see Fig. 2.29 a–c). Thus the greater the silver grain size, the poorer the resolution will become. Hence not only are fine grain emulsions to be recommended for autoradiography, but one tries to produce emulsion layers which are as thin as possible, in order that only those silver bromide crystals lying directly adjacent to the source may be reduced to silver grains. Ideally, these layers are mono-layers of closely packed silver bromide crystals. Unfortunately, such emulsions are much less sensitive than those consisting of several layers of silver bromide crystals.

The smallest silver bromide crystals which are available in an emulsion have a diameter of 60 nm (Kodak NTE-emulsion). With the Ilford L4 emulsion, which is used most often in Europe, the silver bromide crystal size is 140 nm. Thus resolution in EM-autoradiography usually lies between 60 and 140 nm. Since maximal resolution in LM-autoradiography is of the order of 1 μm (1000 nm), this means that resolution in EM-autoradiography is only about ten times better. The advantage of EM-autoradiography is therefore not so much a question of resolution but one of improved allocation of silver grains to organelles which are not recognizable in the LM.

Literature

Bachmann L, Salpeter MM (1967) Absolute sensitivity of electron microscope autoradiography. J Cell Biol 33: 299-305

Caro LC (1962) High-resolution autoradiography II. The problem of resolution. J Cell Biol 15: 189-199

2.5.5.3 *Preparation of Sections*

Sections of embedded radiolabelled material are cut as for normal TEM (see Sect. 2.2.4). Slightly thicker sections (90 nm thick, gold colour) are generally better for autoradiography than very thin (<70 nm thick) ones. Sections are best taken up on formvar-coated grids. Many workers recommend nickel instead of copper grids, since the latter tend to change colour strongly during the development of the emulsion, although this has no effect on the sections.

The type of grid used depends upon the type of method employed for coating the sections with emulsion (see below). For the loop method, 100-200 mesh grids are to be recommended. In contrast, for the run-off method, slot-grids are preferable. Although in this case the formvar film is exposed to considerable forces during the various steps in autoradiography, is is not normally necessary to stabilize the film with carbon. With methods such as the dipping technique, in which sections are laid onto formvar-coated slides and then coated with emulsion, exposed, developed and fixed before mounting onto the grids, the type of grid used is irrelevant. One should note that in those methods where the liquid emulsion comes into contact with a grid, the emulsion tends to be thicker at the junctions between the grid meshwork and the formvar film than elsewhere.

2.5.5.4 *LM Autoradiography*

A knowledge of LM-autoradiography is indispensable for the application of this method at the EM level; indeed a thick section (1 μm) for LM purposes should always be prepared at the start of an EM investigation in order to check the correct exposure time. The necessary chemicals and apparatus for LM auto-

Table 2.13. Chemicals required for light microscope autoradiography

Absolute alcohol	Ether
Chrome alum	Gelatine (capsules)
Photoemulsion	$CaCl_2$ or other drying agent
Kodak-D19-developer	Kodak fixer
Staining solution (e.g. Toluidine blue)	Sealing agent (e.g. Euparal)

Table 2.14. Apparatus required for light microscope autoradiography

Microscope slides	Cover slips
Slide holder	Staining jar(s)
Water bath	Timer
At least five glass beakers (2 × 50 ml; 3 × 100 ml)	
Wire holder to position staining jar in water bath	
Light-tight exposure box(es)	

radiography are given in Tables 2.13 and 2.14. A brief description of the method is as follows:

Thick sections are transferred to glass microscope slides carefully cleaned with a 1:1 mixture of alcohol and ether. Normally five slides with at least three sections are prepared per object. The slides are then coated with a layer of chrome-alum gelatine which enables the photographic emulsion to stick better to the sections. The solution is prepared by dissolving 0.5 g chromium III potassium sulphate dodecahydrate in 1 l distilled H_2O at 37° C. When dissolved, 5 g gelatine is added and stirred until a clear solution is obtained. After being dipped in the chrome-alum gelatine solution, the slides are then held vertically over filter paper to dry. The slides can be processed immediately for autoradiography but can also be stored for several weeks in a dust-free box.

The slides are then coated with emulsion by the "dipping" technique, which is carried out in the dark room under red light. A 50-ml beaker containing 25 ml distilled H_2O and a staining jar for slides are placed in a water bath at 42° C. Ilford L4 emulsion is added to the beaker under stirring until the volume has reached 50 ml. The emulsion solution is poured into the staining jar and the slides dipped into the emulsion. The slides are then carefully removed and allowed to dry (this takes at least 1 h) vertically on filter paper. Alternatively, the slides can be coated via the "run-off" method using the same apparatus described below for EM-autoradiography (see Sect. 2.5.5.5 c) where the slide with thick sections replaces the glass T.

Finally the dry, emulsion-coated slides are placed in a light-tight box which contains some $CaCl_2$ as a drying agent, and left exposed in a refrigerator at 4° C. After about a week one of the slides is removed and developed in a staining jar with Kodak D19 developer for 3 min at room temperature. After washing (by dipping in each of three 100-ml beakers filled with distilled H_2O), the slide is placed in Kodak fixer for 15 min. The slide plus sections can then be stained, e.g. with toluidine blue (see Sect. 2.2.1.2) and, if desired, made permanent by sealing with Euparal or Caedax. If a positive reaction, as judged by the presence of silver grains, is not attained after a 1-week exposure, the above procedure is repeated with another slide a week later etc. until a clear result has been achieved. As a rule the minimum exposure period for EM-autoradiography is at least five times (usually ten) that determined for LM-autoradiography.

Literature

Rogers AW (1979) Techniques of autoradiography. Elsevier, Amsterdam, pp 1-430

2.5.5.5 *Emulsion, Coating Techniques*

As already mentioned, the aim here is to coat the sections with a monolayer of silver bromide crystals. There are numerous methods available, three of which are described here. Of these the "loop" method is the most reliable and allows the production of the most even layers.

a) The "Loop" Method

For this method two small pieces of apparatus are necessary: a holder or support for the grids and a platinum loop. Both are shown in Fig. 2.30. The former is best made out of perspex with the vertical rods having a diameter of around 2 or 3 mm (i.e. similar to that of a grid). From experience the best films are obtained when the loop is rectangular (30 × 15 mm; diameter of wire 0.3-0.5 mm),

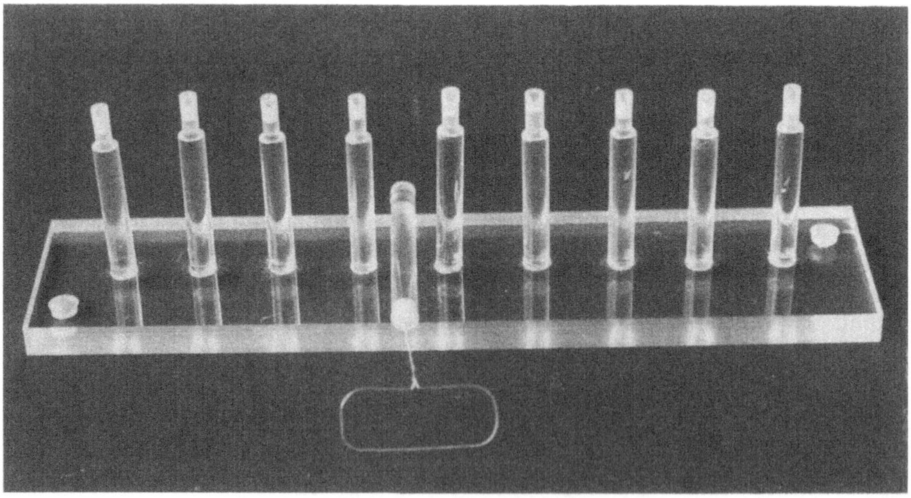

Fig. 2.30. Platinum loop and perspex grid support stand for coating sections with photoemulsion via the loop method

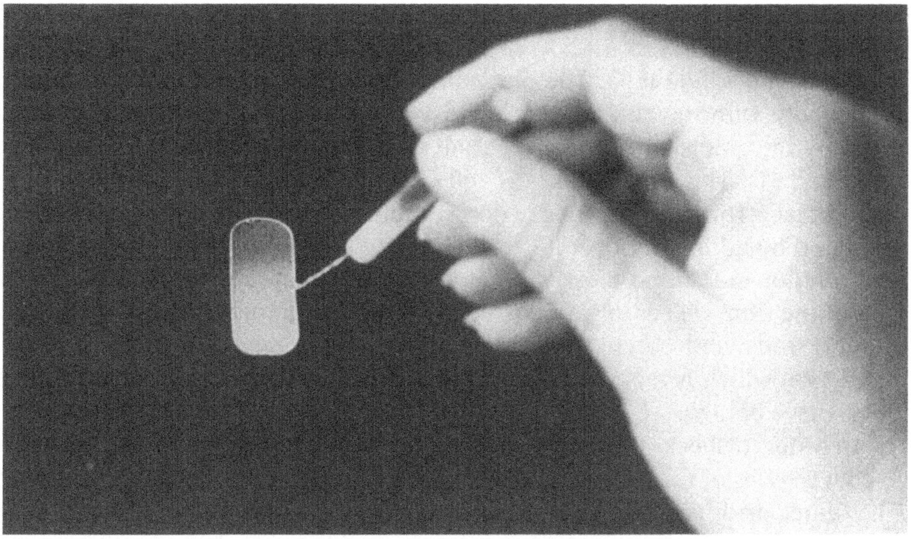

Fig. 2.31. Platinum loop with an emulsion just beginning to gel. The emulsion is thinnest in the uppermost *(darkest)* portion of the film

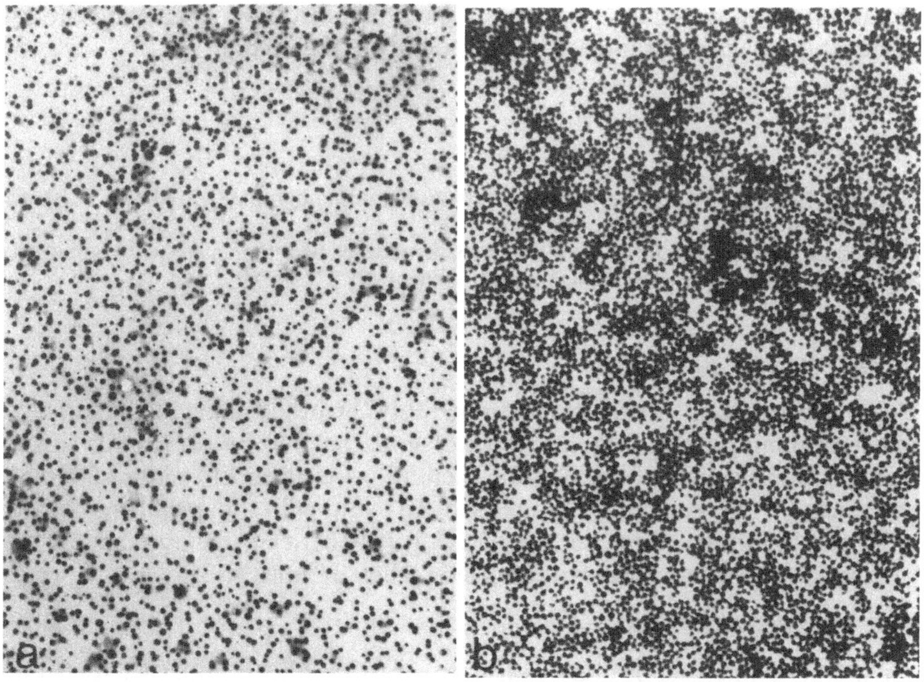

Fig. 2.32. **a** A monolayer of Ilford L4 photoemulsion which has been excessively diluted. **b** Ilford L4 emulsion at the correct dilution (1 part emulsion : 1 part distilled water)

rather than circular in shape. The photoemulsion, Ilford L4, is prepared as described above (see Sect. 2.5.5.4) in the dark room at 42° C but diluted 1 part: 2 parts with distilled H_2O. The platinum loop is then dipped into the emulsion and slowly withdrawn. It is then held vertically until the film begins to gel, whereby the emulsion in the upper half of the loop is generally thin and even (see Fig. 2.31). This usually takes 30-60 s. If the emulsion film in the loop bursts soon after its formation, the emulsion solution is too dilute and should be concentrated by adding more emulsion. The loop is then carefully lowered onto a grid (section-side uppermost) so that it becomes coated with the thinnest portion of the film. The emulsion-coated grids are then removed from the perspex support rods and stuck with their edges onto a strip of sticky tape (see Fig. 2.33 a) which is mounted onto a glass T. In this form they are ready for exposure (see below).

In order to check the density of silver bromide crystals in the film it is useful to coat a series of formvar-coated grids with different dilutions of the emulsion. The filmed grids are then inserted directly in the electron microscope where the density of silver grains can easily be determined (see Fig. 2.32)

The loop method is the method of choice for coating cryosections. Since it is imperative that such sections do not come into contact with water after they are cut, they are best coated with a dry emulsion film. To achieve this one simply allows the film in the loop to dry. Actually this is not as easy as it sounds and the film may well burst; if so it is better to prepare thicker films (more concentrated emulsion solutions) than to risk coating with a moist film, which can lead to a dislocation of soluble radioactive compounds in the section.

Literature

Caro LG, Tubergen RP van (1962) High resolution autoradiography I. Methods. J Cell Biol 15: 173-188

Yoshida K, Murata F, Ohno S, Nagata T (1973) A modified wire-loop method for quantitative electron microscopic radioautography. Histochemistry 57: 93-96

b) The "Dipping" Method

Although the "loop" method is without doubt the best coating method, it is too time-consuming when large numbers of grids have to be processed. In such cases a modification of the dipping method, as used in LM-autoradiography (see Sect. 2.5.5.4), is to be recommended. To avoid adhesion problems between grid and liquid emulsion (see Sect. 2.5.5.3) it is better if the sections are first coated, developed and fixed before being taken up onto grids.

One proceeds as follows:

Coat a glass microscope slide with formvar as described under Chapter 1.4.2. Sections are then transferred with the help of a mounted eyelash onto the formvar film, at best in the middle of the slide. This must be done carefully, transferring at the same time as little water as possible. Excess water is removed with filter paper. When sufficient sections have been mounted the slide can then be dipped into the emulsion solution (prepared as described for LM-autoradiography with an optimal dilution of 1 part emulsion: 6-8 parts dist. H_2O). The slide is slowly drawn out of the emulsion and dried (in darkness) in a vertical position (e.g. by placing in a small beaker) for at least 1 h. The slides with emulsion-coated sections can then be placed in a light-tight box for exposure as described for LM-autoradiography.

Literature

Kopriwa BM (1973) A reliable, standardized method for ultrastructural electron microscopy radiography. Histochemistry 37: 1-17

c) "Run-Off" Method

Also very quick and suitable for large numbers of grids is the "run-off" method. As already mentioned, for this method sections are collected on slot grids to avoid adhesion effects at grid bars. The grids are then stuck by their edges onto a strip of stick tape, which is itself stuck onto a glass T (see Fig. 2.33 a). This T is inserted into a cylindrical dropping funnel of the same sort as that used for coating grids with formvar (see Fig. 2.33 b and also Chap. 1.4.2). However, because the photoemulsion has to be maintained at a temperature of 42° C the coating apparatus has to be inserted into a water bath. Moreover, in order to control the level and speed with which the emulsion enters and leaves the dropping funnel a simple hand-pump assembly is required (see Fig. 2.34 a, b).

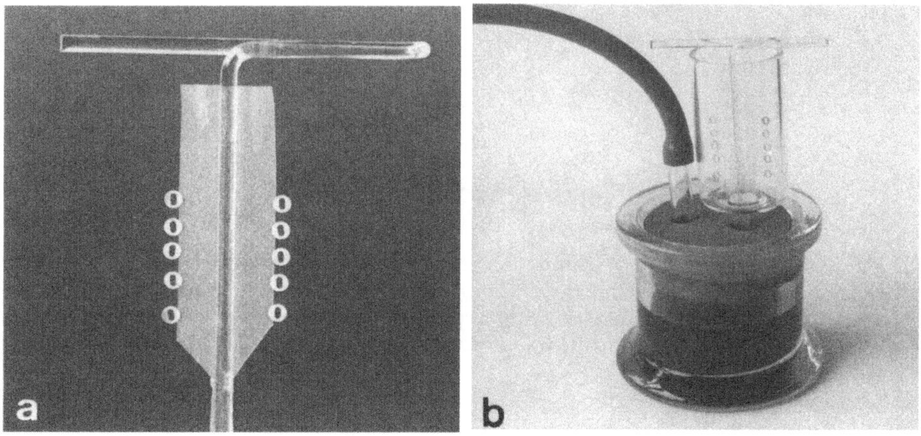

Fig. 2.33. **a** Glass T with a piece of sticky tape to which slot grids are attached. **b** Glass T positioned in a dropping funnel which is inserted in a glass holder as part of the apparatus required for coating with emulsion via the "run-off" method

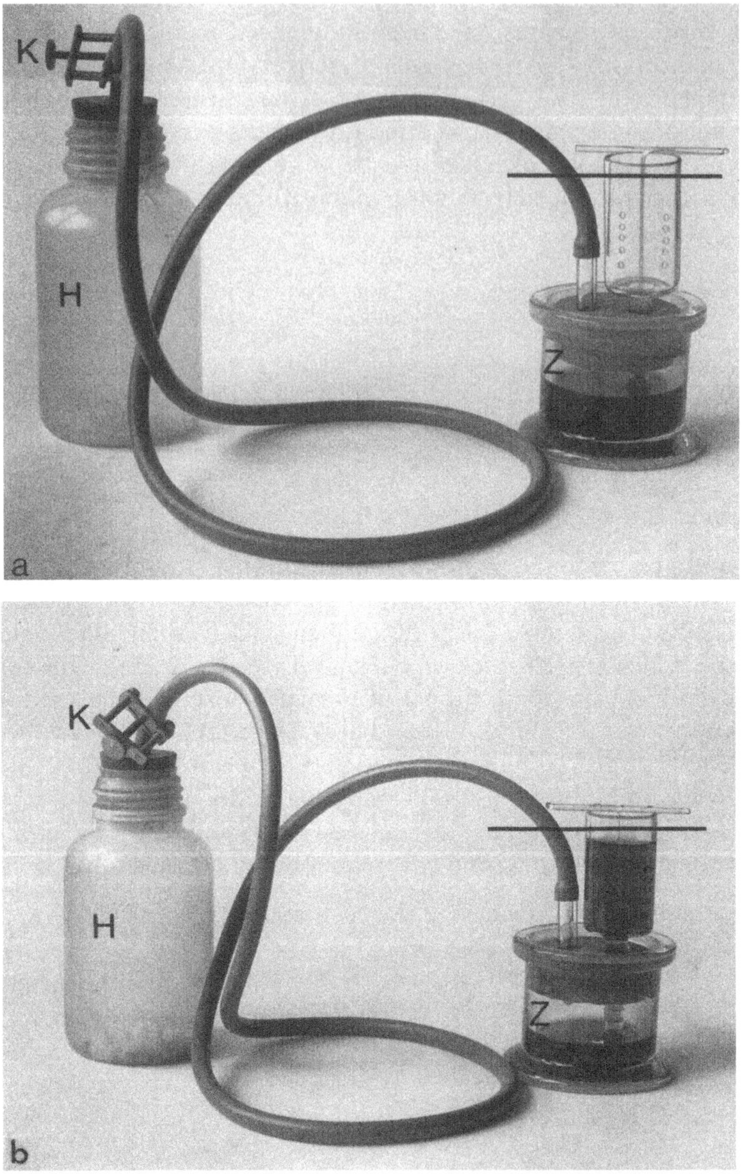

Fig. 2.34 a, b. Apparatus for coating with emulsions via the "run-off" method. **a** Before the emulsion has been pumped into the dropping funnel. **b** Showing the emulsion in the dropping funnel. *H* hand pump; *K* clamp; *Z* cylindrical glass holder. The glass and the dropping funnel are maintained in a water bath, the water being indicated by a *black line* in each photograph

Photoemulsion (Ilford L4; diluted 1 part:4 parts dist. H_2O) is pumped into the dropping funnel. Excess emulsion is allowed to run off the grids by leaving the glass T in the funnel for 1-2 min. The glass T plus coated grids is then hung in a light-tight box (see below) for exposure.

As with the other coating methods, a series of test coatings with formvar-coated grids should be carried out in advance. One should note that not only the relative dilution of the emulsion is important here, but also the speed (as is the case in the production of formvar films; see Chap. 1.4.2) with which the emulsion runs out of the dropping funnel.

2.5.6 Exposing, Developing and Fixing

2.5.6.1 Exposing

As with exposures in LM-autoradiography, emulsion-coated EM specimens must be held in a dry, light-tight box at 4° C. Exposure boxes can easily be made in a workshop according to local needs. A useful construction made of PVC for exposing grids stuck to a glass T is shown in Fig. 2.35; a drying substance such as $CaCl_2$ powder can be added in a small muslin bag in one corner.

There are no rules as to the correct duration of an exposure. As already mentioned, a guide to determining the appropriate exposure time is given from LM sections (see Sect. 2.5.5.4). Thus when a LM section gives an adequate ex-

a b

Fig. 2.35. a PVC holder for glass Ts. **b** Light-tight box used for exposing emulsion-coated sections. The PVC holder with glass Ts fits into this box

posure in 14 days, the corresponding EM section will require 5 months. Even if this period can possibly be made shorter by using radioactive compounds of higher specific activity, one can rarely achieve exposure times in EM autoradiography of less than 2-3 months.

With such long exposure periods the chances of unspecific reduction of silver bromide crystals in the emulsion naturally increase. One must therefore pay attention to all theoretically possible sources of contamination. Thus all traces of detergent should be removed from the exposure box by rinsing in distilled H_2O. Even faulty contacts in thermostats in refrigerators can lead to an unspecific reduction in the photoemulsion. The light-tight boxes should not be touched during the exposition period. If possible, those specimens which are removed during the exposure period in order to control the extent of silver grain development should be kept separate from those which serve to provide the final results. The refrigerator in which the boxes are held should be used solely for this purpose and opened only when absolutely necessary.

2.5.6.2 Developing and Fixing

Two methods are available for developing radioactively exposed photoemulsions: a chemical procedure which results in the production of randomly coiled threads of silver grains, and a physical fixation which results in the formation of silver grains smaller than the original silver bromide cystals (20-60 nm diameter as opposed to 0.14 μm, the usual crystal size in Ilford L4 emulsion). In the latter case this does not lead to an improvement in resolution, since this is, as already discussed (see Sect. 2.5.5.2), dependent on the silver bromide crystal size. (If, however, after development, silver grains are formed which are larger than the original crystals, these determine the resolution instead). Of the two methods, only the chemical fixation is used for autoradiographic purposes.

Ideally, with the best developer the size of the silver grain should not differ from that of the original silver bromide crystal. This is usually achieved with the so-called compact developer, which is prepared by dissolving

1.5 g ascorbic acid
0.25 g Phenidone (Geigy or Tetenal)
0.6 g potassium bromide
1.3 g potassium carbonate
3.0 g potassium rhodanide

one after another in 75 ml distilled water and then diluting to 100 ml. The solution should always be freshly made up and filtered before use. Two other chemical developing systems are in use. In the one case 1.125 g Amidol (diaminophenal), 4.5 g anhydrous sodium sulfite and 2 ml of 10% (w/v) potassium bromide solution are dissolved in 250 ml distilled water. This is the so-called Amidol developer. A ready-to-use developer, the Kodak-D19 developer, is also available commercially. If the developer is too concentrated, the silver filaments may not lie compactly. This can be corrected by suitable dilution with water. Since resolution is more or less the same with all three types of developer, the choice of a developer is generally based on personal tastes and financial arguments.

Developing grids which are stuck on sticky tape proceeds just like any other photographic development: the glass T is lowered gently into the developer and left there (in the dark) for 3 min at room temperature (20° -22° C). Longer developing times may be necessary, but there is always the risk of unspecific silver bromide reduction with increasing developing times. The developing process is stopped by transferring the glass T with grids to a water bath (some workers add acetic acid - end concentration 2%, v/v - to the stopper bath, but this is not absolutely necessary). After 10-15 s the glass T and grids are put into fixer, where they remain for 10 min. A normal negative film fixer is fully adequate for this purpose. Finally, the specimens are rinsed in water by transferring the glass T plus grids several times into beakers with fresh distilled water (gentle agitation helps the rinsing and the specimens are left 2-3 min in each change of water). After being washed, the specimens can be removed with tweezers and stained in the manner as described under Section 2.2.6.

In the dipping method, in contrast to the loop and run-off methods, sections and emulsion are mounted onto grids after developing and fixing the emulsion layer. Before this is done, however, it is usual to stain the sections with heavy metal salt solutions whilst they are still on the glass slide. After staining, the sections can be removed from the slide by floating the entire formvar film onto water exactly as described under Chapter 1.4.2. The sections with overlying emulsion now lie on the formvar film. Grids are then placed carefully onto the sections and the formvar film removed from the water surface onto paper.

When viewing autoradiographic preparations in the EM it is to be recommended that observations and exposures should be carried out as quickly as possible. By extended observation periods or with too high a beam intensity it is possible for silver grains to become separated from the emulsion.

2.5.6.3 Future Developments in Autoradiography

There are three areas in autoradiography where possible technical improvements might be wished for. The first concerns resolution, which is poor considering the overall resolving power of the EM. At the moment, however, it is not possible to make emulsions with silver bromide crystals smaller than 60 nm, and it looks as if there will be no improvement in this direction.

The second area concerns the long exposure periods in EM-autoradiography. According to some authors, exposition times can be reduced eight- to tenfold by coating the sections on both sides with an emulsion and by over-developing the emulsion (6 min instead of 3). Another way to shorten exposure times is to use scintillation solutions. This is termed fluorography and involves exchanging the emitted electrons for photons. This certainly has advantages in the detection of ^3H in sections which are thicker than 3-5 μm. Since ^3H radiation extends only 3-4 μm on the average, the majority will not leave such thick sections. Photons, however, have a much greater radius in which they can be detected. However, three electrons are required to cause the release of one photon, thus only one third of the radioactivity in the specimen can be visualized in this way. Moreover, because of the greater radius of detection of photons, the resolution decreases markedly. These two disadvantages, together with the fact that sections are much thinner than the radius in which electrons from tritium can be detected, do not make fluorography an interesting prospect for EM-autoradiography. Nevertheless, some workers have claimed to have achieved a reduction in exposure time from 3 months to 1 week through the use of a scintillant.

The third area concerns the detection and localization of water-soluble radioactivity. Two methods exist (freeze drying followed by vacuum embedding, and cryo-ultramicrotomy), by which such substances can be processed for EM-autoradiography. Unfortunately, there are few workers who are proficient in these methods and the necessary apparatus is not yet universally available in ultra-structure laboratories. This should change in the years to come.

Literature

Gregg M, Reznik-Schüller HM (1984) An improved method for rapid electron microscopic autoradiography. J Microsc 135: 115-118

Mizuhira V, Shiihashi M, Fugicka T (1979) High speed autoradiography in electron microscopy. Acta Histochem Cytochem 12: 490

2.6 Freeze (Fracturing) Etching

2.6.1 Introduction

Freeze etching offers an alternative to conventional methods of preparation. Instead of a chemical fixation with subsequent dehydration, embedding and thin sectioning, this technique involves a physical fixation: freeze fixation. The replicas which are finally produced contain much more information on the structure of biomembranes and other cell and tissue components than do thin sections.

The basic principles of this method were laid down in the 1950's, making it just as "old" as, for example, aldehyde fixation or Epon embedding. However, because of the relatively high costs involved, the freeze-etching apparatus did not become a regular piece of equipment in cytological laboratories until the middle of the 1970's.

The freeze-etching method can be divided into five steps, the first of which (freezing) and the last (replica cleaning) are performed without a freeze-etching apparatus. Of the other three steps, etching (vacuum sublimation) is not obligatory: if omitted, the technique is then termed freeze fracturing.

Literature

Mühlethaler K (1973) History of freeze-etching. In: Benedetti EL, Farvard P (eds) Freeze-etching: techniques and applications. Soc Fr Microsc Electron, Paris, pp 1-10
Rash JE, CS Hudson (eds) Freeze fracture: methods, artifacts and interpretations. Raven, New York, pp 1-204
Robards AW, UB Sleytr (1986) Low temperature methods in biological electron microscopy. In: Practical methods in electron microscopy. vol X Elsevier/North-Holland, Amsterdam pp 1551
Wilson JHM, AJ Rowe (1980) Replica, shadowing and freeze-etching techniques. In: Glauert AM (ed) Practical methods in electron microscopy, vol III. Elsevier/North-Holland, Amsterdam, p 1301

2.6.2 Freezing

2.6.2.1 Theoretical Background

In freezing biological material one must try to keep the damage caused to organelles by ice crystal formation down to a minimum. Two phases are created when a solution (the cytosol for example) is frozen: ice crystals and the so-called eutectic, which comprises water and the various dissolved substances. The latter is solid but amorphous ("vitreous"). In order to prevent ice-crystal formation, one freezes as quickly as possible in order to reach the vitreous state.

Both the temperature which is finally reached as well as the freezing rate determine the relative success of a freeze fixation. Ice crystals are formed in pure water between temperatures of 0° C and −130° C; if solutes are present the latter temperature (the recrystallisation point) is raised. In the case of cytoplasm this lies around −80° C, which means that a freezing agent with a temperature of at least −100° C must be employed.

The freezing rate determines the number and type of ice crystals formed. At low freezing rates, both solutes as well as the water molecules themselves can serve as nuclei for ice-crystal formation. At higher freezing rates, only the latter are involved in this process, resulting in the production of large numbers of small, homogeneous crystals. The change from heterogeneous to homogeneous ice crystal formation occurs suddenly when the so-called critical freeze rate is reached or overstepped. For pure water this parameter is almost 10^{6}° C s^{-1} but for cytoplasm it lies around 10^{4}° C s^{-1}. These values relate to layers of only a few μm in thickness and to spherical objects of up to 20 μm in diameter. Larger objects can only be effectively frozen in the vitreous state when other physical factors are changed. One such possibility is the hydrostatic pressure. Thus the critical freezing rate for H_2O can be reduced by 10^{2}° C s^{-1} when the pressure is raised to 2000 bar; however, in order to avoid cellular damage, such pressures can only be maintained for a fraction of a second. A commercial high-pressure freezing apparatus is available from Balzers. Another possibility is the use of a cryoprotectant (see below).

Literature

Moor H (1964) Die Gefrierfixation lebender Zellen und ihre Anwendung in der Elektronenmikroskopie. Z Zellforsch 62: 546-580
Moor H (1973) Cryotechnology for the strucural analysis of biological material. In: Benedetti EL, Favard P (eds) Freeze-etching techniques and applications. Soc Fr Microsc Electron, Paris, pp 11-19
Moor H, Bellin G, Sandri C, Akert K (1980) The influence of high pressure freezing on mammalian nerve tissue. Cell Tissue Res 209: 201-216
Nei T (1973) Growth of ice crystals in frozen specimens. J Microsc 99: 227-233
Sitte H (1979) Cryofixation of biological material without pretreatment - a review. Mikroskopie 35: 14-20

2.6.2.2 Cryoprotectants

The critical freezing rate can be reduced considerably by the use of a suitable cryoprotectant. For example, cytoplasm impregnated with 20% glycerine has a critical freezing rate of only $10^{2\circ}$ C s^{-1}. In addition, as a result of the presence of a large number of osmotically active particles, the freezing point is lowered and the recrystallisation point is raised. As a result, the temperature range in which ice crystals can form is considerably reduced.

One normally differentiates between permeating and nonpermeating cryoprotectants. Well-known representatives of the first group are glycerine and dimethyl sulfoxide (DMSO). Usually these substances are given to cells in concentrations increased stepwise as buffered solutions at 4° C. For animal cells exposure to a 20% (v/v) solution for 90 min is usually quite adequate. One must be careful in using these types of cryoprotectant: not only can they be a health hazard, they may also induce a number of artefacts in the material being investigated:

- a segregation of integral membrane protein (IMPs) through a redistribution of membrane lipids,
- the vesiculation of endomembranes,
- plasmolysis in plants.

Because of these problems, a stabilization with glutaraldehyde (1%, v/v, 1 h; "milder" than a normal fixation; see Sect. 2.1.1) prior to the freeze fixation is often carried out. One should, however, note that although many of the above-mentioned artefacts can thus be avoided, such a pretreatment is not in itself free of artefact. A number of changes in membrane structure are known to occur as a result of a glutaraldehyde fixation which do not make the combined use of aldehydes and cryoprotectants a method of first choice.

Nonpermeating polymers such as polyvinyl pyrrolidone, starch and dextran have not enjoyed much success as cryoprotectans. Although their mode of action is unclear, only short exposures to them are necessary. On the other hand, it has been reported that their use can lead to changes in the fracture plane in membranes. As yet they have not been used much on animal cells.

Literature

Franks F (1977) Biological freezing and cryofixation. J Microsc 119: 3–10
Plattner H, Bachmann L (1982) Cryofixation: a tool in biological ultrastructural research. Int Rev Cytol 79: 237–304
Richter H (1968) Die Reaktion hochpermeabler Pflanzenzellen auf drei Gefrierschutzstoffe (Glyzerin, Ethylenglykol, Dimethylsulfoxid). Protoplasma 65: 155–166
Skaer H le B, Franks F, Asquith MH, Echlin PH (1977) Polymeric cryoprotectants in the preservation of biological ultrastructure III. Morphological aspects. J Microsc 110: 257–270

2.6.2.3 ***Supports***

Because of their good thermal conductivity, gold and copper are used for freeze-etch specimen supports. The simplest form is that of a 3 × 0,5 mm disc, which is scratched in the middle to enable better attachment of the object (Fig. 2.36a). For tissue (pieces of 0.5 mm^3 or smaller) a specimen support with a

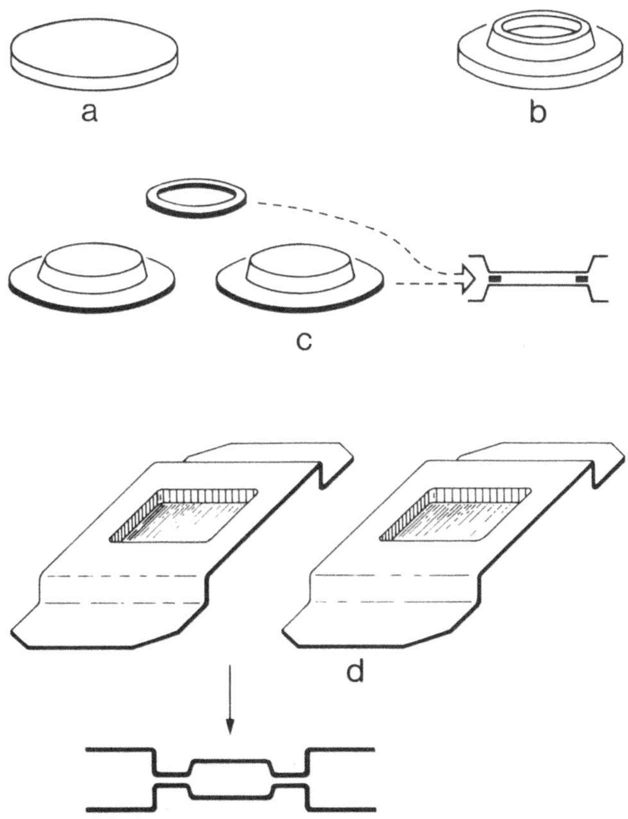

Fig. 2.36 a–d. Specimen supports for freeze etching: **a** simple (disc); **b** disc form with raised annulus (both 3 mm in diameter); **c, d** supports for complementary replica devices

raised central annulus (Fig. 2.36b) is often used. The material is mostly stuck into the support disc with the help of gum arabicum (4%, w/v) or with a thick paste of yeast cells. These supports are relatively large (about 50 mg) and should only be used for conventional freeze fixations (see Sect. 2.6.2.4).

For the ultra-quick freezing methods, much smaller (about 10 mg) supports are in use. These can be obtained commercially (Balzers, Fig. 2.36d) or can be made in the laboratory from thin (0.1 mm) copper plate with the help of a suitable punch. Usually, thin supports are used pair-wise in one of the various "sandwich" freezing procedures. An evently spread cell or organelle suspension (usually 1-3 μl) between the supports can be achieved by employing ring-like spacers (Fig. 2.36c), which are punched out of 3-mm diameter EM grids.

Both spacer and supports should be cleaned before use by sonicating 2-5 min in a mixture of absolute methanol and 1N HCL (20:1, v/v), rinsing in dist. H_2O and drying on filter paper. Because of the dangers of oxygen deficiency and/or dehydration, specimens should be mounted and frozen individually.

Literature

Gulik-Krzywicki T, Costello MI (1978) The use of low temperature X-ray diffraction to evaluate freezing methods used in freeze-fracture electron microscopy. J Microsc 1112: 103-114

Müller W, Pscheid P (1978) A new inexpensive specimen carrier for freeze-fracturing. J Microsc 115: 113-116

Staehelin LA (1982) Freeze-fracture and freeze-etch electron microscopy of chloroplast membranes. In: Edelman M, Hallick RB, Chua N-N (eds) Methods in chloroplast molecular biology. Elsevier/North-Holland Biomedical Press, Amsterdam, pp 821-833

2.6.2.4 Cryogens and Freezing Methods

Only liquefied gases which do not show the Leidenfrost phenomenon (reduced cooling rate due to the production of a protective gas layer) can be employed as cryogens. Liquid nitrogen, therefore, cannot be used. Instead, either Freon 12 (dichloro difluoromethane, melting point −159° C), Freon 22 (monochloro difluoromethane, melting point −160° C) or propane (melting point −187° C) are in use. One should note that, because of its explosive nature, propane should only be used in a fume cupboard.

In the so-called dip methods the cryogen gas is directed from a cylinder into a metal holder which is pre-cooled with liquid N_2 (Fig. 2.37A). The liquified cryogen solidifies very quickly and therefore has to be rewarmed before use. This simply involves pressing a metal rod into the solidified cryogen. Single or sandwich-type supports held with fine tweezers are then inserted into the liquified cryogen. After several seconds the frozen specimens are removed and quickly transferred to liquid N_2. The freezing rate with this method lies at best around 10^{3}° C s^{-1} for the larger supports, making ice-crystal formation almost unavoidable in the absence of a cryoprotectant. The plasma membrane, nevertheless, usually remains well preserved. In contrast, sandwich-mounted speci-

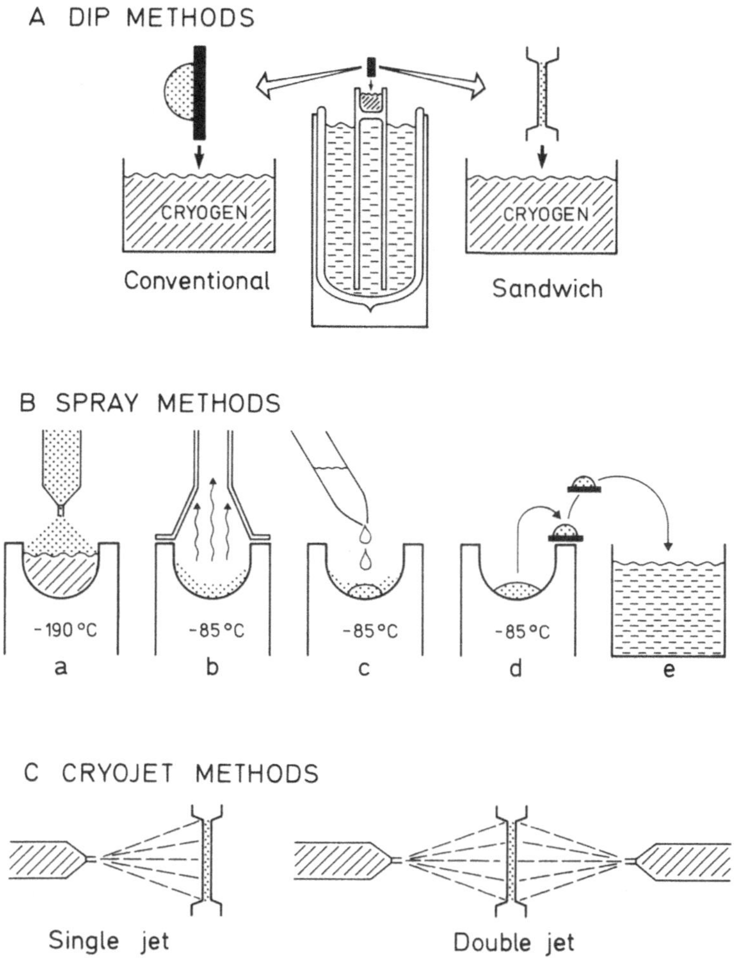

Fig. 2.37 A-C. Freezing methods for the freeze-etch technique (for details see text)

mens with thin supports can be frozen somewhat faster (between 10^3 and $10^{4\circ}$ C s^{-1}).

Higher freezing rates can only be achieved through one of the "cryojet" methods (Fig. 2.37 C). The liquified cryogen, usually propane, is "shot" with the aid of a compressed air pistol (operated with N_2 gas, 4 bar) at a sandwich-mounted specimen placed about 1 cm away. The object can be sprayed uni- or bilaterally. Equipment for this method is available commercially, e.g. from Balzers.

Instead of spraying the cryogen onto the specimen, the specimen, provided it is a suspension of cells or organelles, can be sprayed into the liquified cryogen (Fig. 2.37B). Once again, a compressed air pistol is used to propel fine drops (up to 30 μm in diameter) into the cryogen. Larger objects, e.g. ciliates, can be "shot" out of a syringe fitted with a suitable orifice (an EM aperture can be used for this purpose). After the specimen is frozen, the cryogen is evaporated under vacuum at higher temperatures (−85° C). Precooled n-butylbenzene or toluene is then added dropwise. This builds a thick paste in which the frozen cellular material is suspended. Small portions of this cell-paste are then transferred, using a precooled mounted needle, onto a support maintained at −85° C. The frozen specimens can then be maintained in liquid N_2. An apparatus for this purpose is available from Balzers; however, simpler versions which are just as effective are easy to construct.

2.6.2.5 Storage of Frozen Specimens

Specimens, irrespective of the way in which they are frozen, can be stored more or less indefinitively in liquid N_2. They can be kept in a small Dewar flask fitted with a metal holder if they are to be processed on the same day they are frozen. For longer storage periods special holders which fit into large liquid N_2 containers can be bought or made.

Literature

Bachmann L, Schmitt-Fumian WW (1973) Spray-freezing and freeze-etching. In: Benedetti EL, Farvard P (eds) Freeze-etching: techniques and applications. Soc Fr Micros Electron, Paris, pp 73-79

Knoll G, Oebel G, Plattner H (1982) A simple sandwich-cryogen-jet procedure with high cooling rates for cryofixation of biological materials in the native state. Protoplasma 111: 161-176

Krah S, Staehelin LA, Nettesheim E (1973) A new type of storage container for freeze-etch specimens. J Microsc 99: 349-352

Lang RDA, Corsby P, Robards AW (1976) An inexpensive spray-freezing unit for preparing specimens for freeze-etching. J Microsc 108: 101-104

Müller M, Meister N, Moor H (1980) Freezing in a propane jet and its application in freeze-fracturing. Mikroskopie 36: 129-140

Plattner H, Schmitt-Fumian WW, Bachmann L (1973) Cryofixation of single cells by freeze-spraying. In: Benedetti EL, Farvard P (eds) Freeze-etching: techniques and applications. Soc Fr Microsc Electron, Paris, pp 81-100

2.6.3 Fracturing

2.6.3.1 Transfer of the Object into the Vacuum Recipient

Frozen specimens are transferred to a vacuum recipient in order that the fractured surface may be replicated without becoming contaminated by condensed water. This does not mean that the fracturing step which precedes this replication is oligatorily performed in vacuo. In the Bullivant-Ames method, for instance, frozen specimens are maintained under liquid N_2 and are fractured with

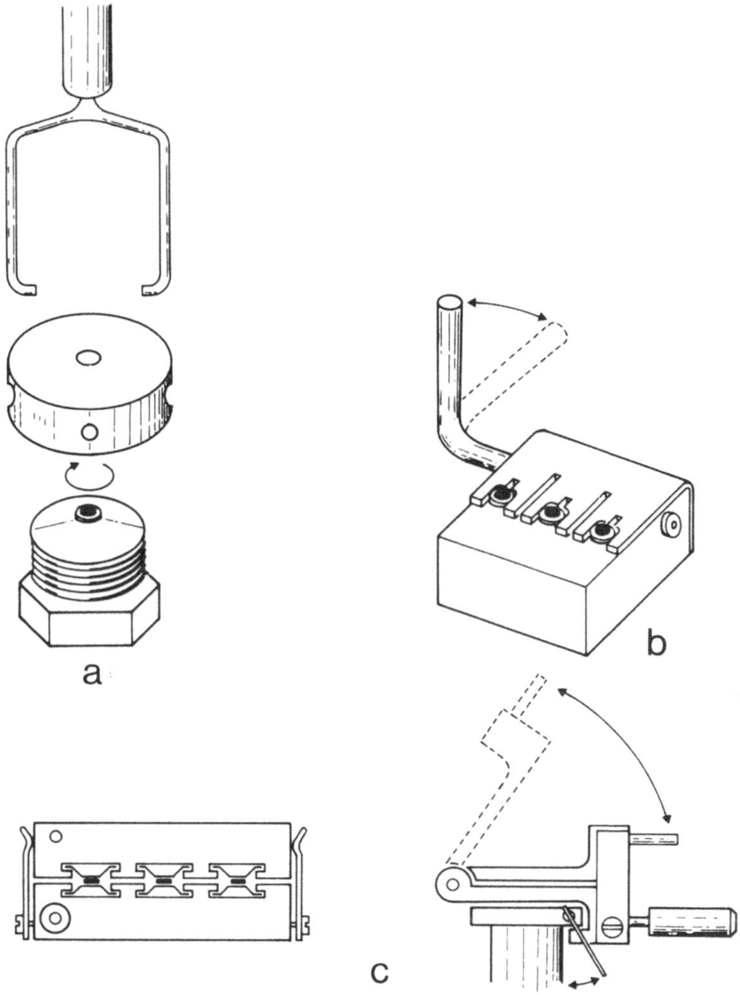

Fig. 2.38 a–c. Specimen tables and object holders used in Balzers freeze-etch machines. **a** Conventional single disc table; **b** multi-disc table; **c** complementary replica device

a scalpel in this "submersed" condition. Only afterwards is the brass fracturing device transferred to an evaporation unit. Replication then occurs after the liquid N_2, which first freezes, is pumped away. Freeze-etch machines built according to this method have the advantage of being cheap, but fracturing and etching tend to be less controllable than in the more conventional type of freeze etcher.

With the other types of freeze-etching apparatus (e.g. Balzers, Edwards High Vacuum), fracturing is carried out in the vacuum recipient. Accordingly, these machines have rotary, oil diffusion or turbomolecular pumps and are equipped with a specimen table and knife assembly, both of which can be cooled by pumping liquid N_2 in and out of them without breaking the vacuum.

In order to reduce the possibility of ice crystal formation, the frozen objects must be transferred as quickly as possible to the precooled (−150° C) specimen table. This also keeps the amount of condensed water on the table to a minimum and enables high vacuum conditions to be reached more quickly. This latter problem is almost completely eliminated in those freeze-etch machines equipped with an airlock through which cold H_2O-free N_2 gas is pumped. Insertion of the specimens into the vacuum recipient occurs against a stream of N_2 gas.

There are two ways in which frozen objects may be transferred to the specimen table. With single, conventionally frozen objects, the supports are transferred individually with fine tweezers and then either clamped or screwed into position with the appropriate holder (see Fig. 2.38 a, b). In this transfer procedure the specimen table is "painted" with liquid Freon prior to mounting the supports in order to provide for a better thermal contact. Objects frozen in the sandwich form are inserted into a special complementary replica holder which is submerged in liquid N_2. After loading, the holder is transferred to the precooled specimen table with the help of special forceps or by gripping an insulated peg, which is part of the holder (Fig. 2.38 c), between finger and thumb.

After loading the specimen table, the recipient is immediately placed under vacuum. A vacuum of at least 5×10^{-5} torr must be reached before fracturing can take place. There are freeze-etch machines (Balzers) available in which an ultra-high vacuum (10^{-8}–10^{-9} torr) can be achieved.

2.6.3.2 The Fracturing Process

Depending on the type of support used, single or double (complementary) fracture surfaces can be prepared. In the former case a knife assembly, precooled to −190° C and fitted with a single-edged razor blade, is employed. Both mechanical and thermal advances are present, although the latter can usually be dispensed with. In order to prevent a splitting or complete loss of the specimen, slow cutting movements with a small advance (1–2 μm) are to be recommended. However, particularly with the last cut, care must be taken to prevent knife marks or scratches appearing on the fracture surface. When properly prepared, a fractured object has a plateau form with a flat, even surface. After fracturing, the knife assembly is either positioned above the object, if etching is to be carried out (see Sect. 2.6.4) or removed away from the object without touching it. Its final position is unimportant, but one has to make sure that neither the evaporating guns nor the thin-film monitor are occluded by it.

The double replica holder for sandwich-mounted specimens is much easier to operate. It is normally a hinged device which is opened either entirely mechanically from outside the vacuum recipient or is partially electrically activated. Although the fracture surfaces which arise are complementary, they need not be treated as such: often a large number of replicas are required. However, if complementarity is desired, some kind of orientation marker must be present in the fracture plane. There are two possibilities here: either one uses two exactly superimposed index grids which are stuck together with Vaseline, or one uses a single index grid whose surface has been coated with lecithin (prepared by dipping into a 0.5%, w/v, lecithin-chloroform solution; excess liquid is removed with filter paper). The grids (or grid) are positioned in the cell or organelle suspension between the two specimen supports and frozen together with them. When the specimen supports are pulled apart, the fracture plane should run between the two grids or within the hydrphobic lecithin bilayer. In this way the letters or numbers in the grids are imprinted in the fracture face and become replicated. They are later recognizable in the EM.

Fracture surfaces which are to be etched are usually made with a specimen table temperature of − 100° C. If there is to be no etching, fracturing is better performed at − 110° C. Because of the danger of surface contamination, fracturing at lower temperatures should only be carried out under high vacuum conditions.

2.6.3.3 Fracture Planes in Biological Material

As a rule, fracture planes run between the two lipid layers of a biomembrane. This represents the hydrophobic region of the membrane where the forces holding the two bilayers together are less than those at the surface of the membrane. In micrographs the fracture planes are prefixed with either a P or an E to denote that either protoplasma or an extraprotoplasmic component lies behind

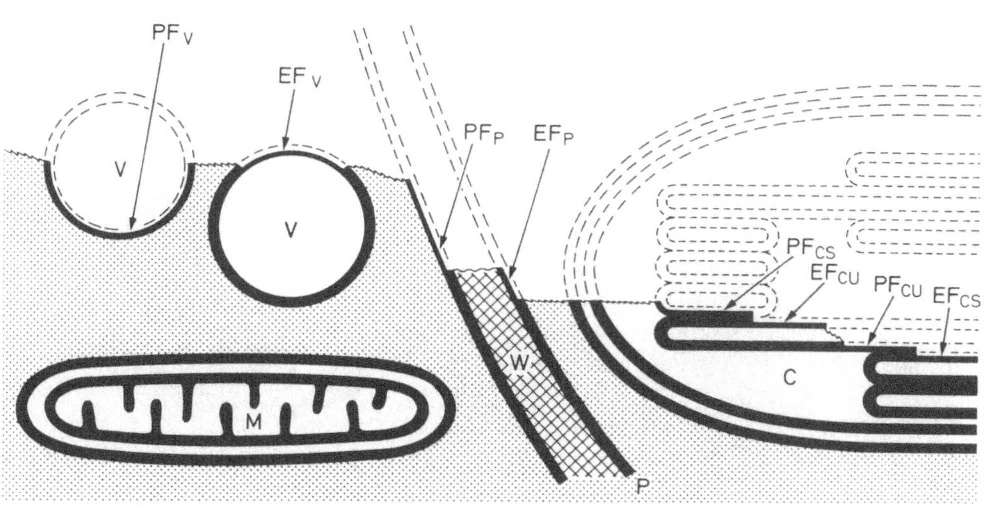

Fig. 2.39. Diagrammatic representation of the passage of a fracture face through two adjacent plant cells. *PF* protoplasmic fracture faces; *EF* exoplasmic fracture faces; *W* cell wall; *P* plasma membrane; *C* chloroplast; *V* vacuole; *M* mitochondrium; *CS/CU* stacked or unstacked chloroplast thylakoids

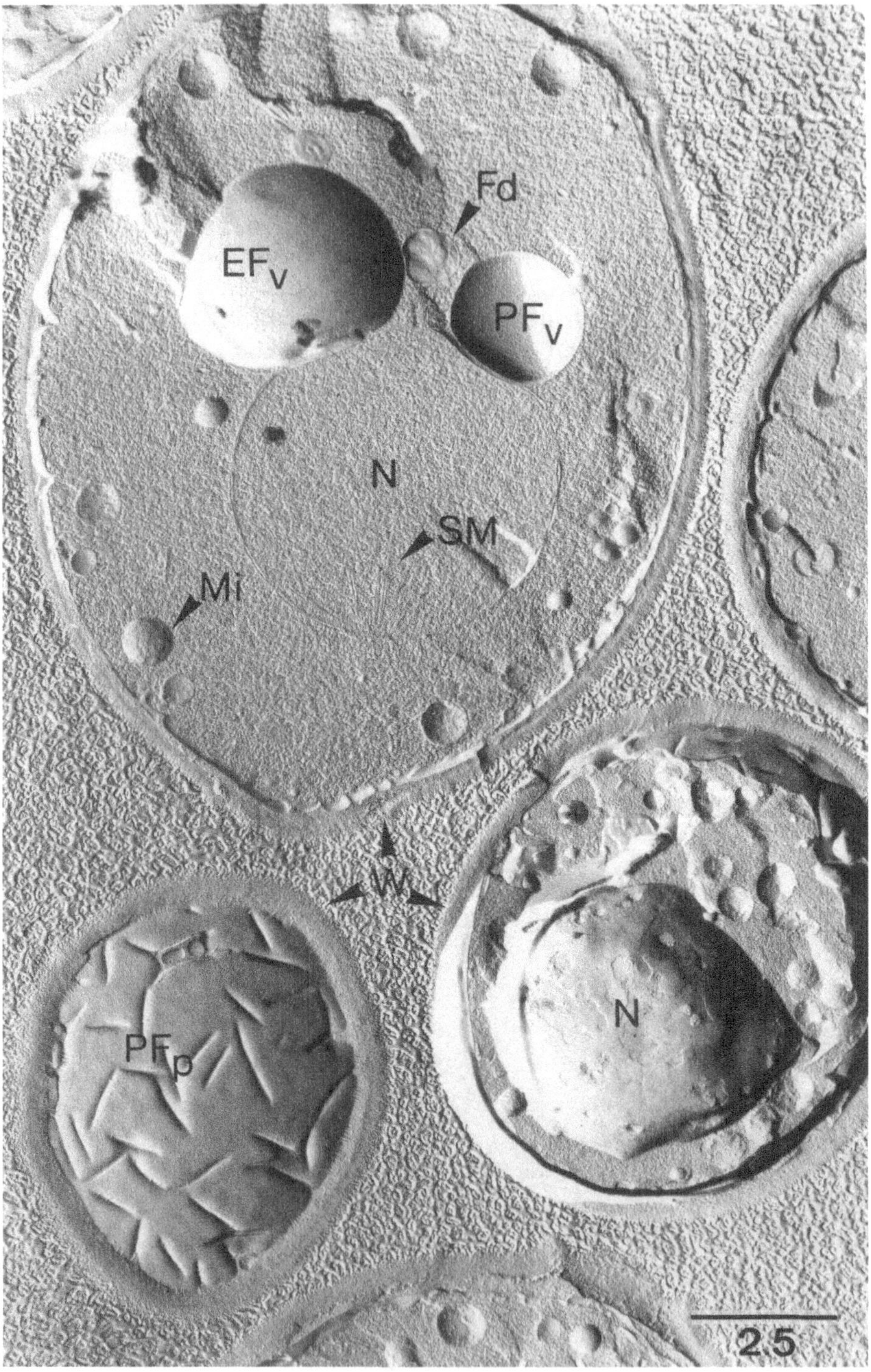

Fig. 2.40

the exposed monolayers respectively. More correctly P indicates the cytosol or the stroma of plastids and mitochondria and E the lumina of the various endomembranes as well as the extracellular space. Fig. 2.39 shows how such fracture planes may arise and Fig. 2.38 is an example of how they appear in an electron micrograph. With few exceptions, the fracture planes of membranes are characterized by particles which probably represent integral membrane proteins (IMPs).

Literature

Branton D, Bullivant S, Gilula NB, Karnovsky MJ, Moor H, Mühlethaler K, Northcote DH, Packer L, Satir B, Satir P, Spetz V, Staehelin LA, Steere RL, Weinstein RS (1975) Freeze-etching nomenclature. Science 190: 54-56

Bullivant S, Ames A (1966) A simple freeze-fracture replication method for electron microscopy. J Cell Biol 29: 435-447

Gross H, Bas E, Moor H (1978) Freeze-etching in ultrahigh vacuum at −196° C. J Cell Biol 76: 712-728

Mühlethaler K, Hauenstein W, Moor H (1973) Double fracturing methods for freeze-etching. In: Benedetti EL, Favard P (eds) Freeze-etching: technique and applications. Soc Fr Microsc Electron, Paris, pp 73-79

Pinto da Silva P, Branton D (1972) Membrane intercalated particles. The plasma membrane as a planar fluid domain. Chem Phys Lipids 8: 265-278

Sleytr UB, Umrath W (1974) A simple fracturing device for obtaining complementary replicas of freeze-fractured and freeze-etched suspensions and tissue fragments. J Microsc 101: 177-186

2.6.4 Etching

2.6.4.1 The Purpose of Etching

Because the fracture plane runs within the lipid bilayer, the true surface of a biomembrane remains occluded in a freeze-fracture preparation. In order to reveal such surfaces the surrounding eutectic has to be removed. This is achieved by etching, whereby exoplasmic or protoplasmic surfaces become exposed (Figs. 2.41 and 2.42).

2.6.4.2 Theory and Practice

Etching is a superficial freeze drying which is performed immediately after fracturing. Under vacuum frozen water sublimes from the fracture face, pro-

◁

Fig. 2.40. Micrograph of a freeze-fracture replica of cells of baker's yeast (*Saccharomyces* sp.). The nomenclature of the fracture faces is that of Fig. 2.37. *Fd* fat droplet; *N* nucleus (the nucleus in the cell at the top of the picture is engaged in mitosis); *Mi* mitochondria; *SM* spindle microtubules; *W* cell wall. Magnification bar in μm

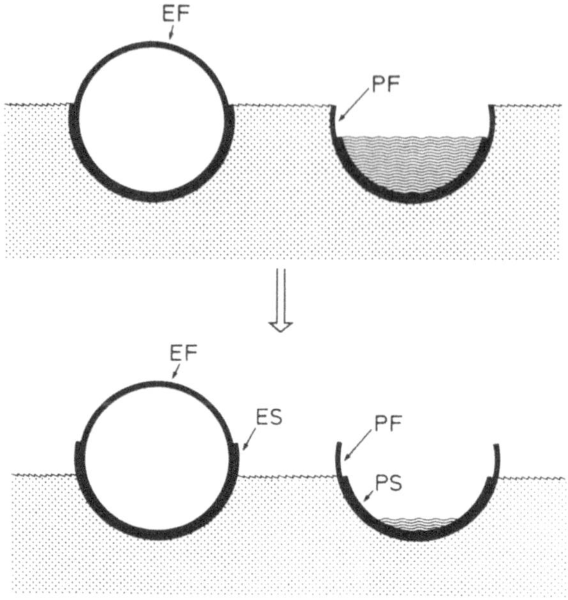

Fig. 2.41. Diagrammatic representation of the principle of etching as exemplified on a vacuole. *Above:* after freeze fracturing; *below:* after a subsequent etching. *ES/PS* "exoplasmic" surface/"protoplasmic" surface. Ice is indicated by *hatching*

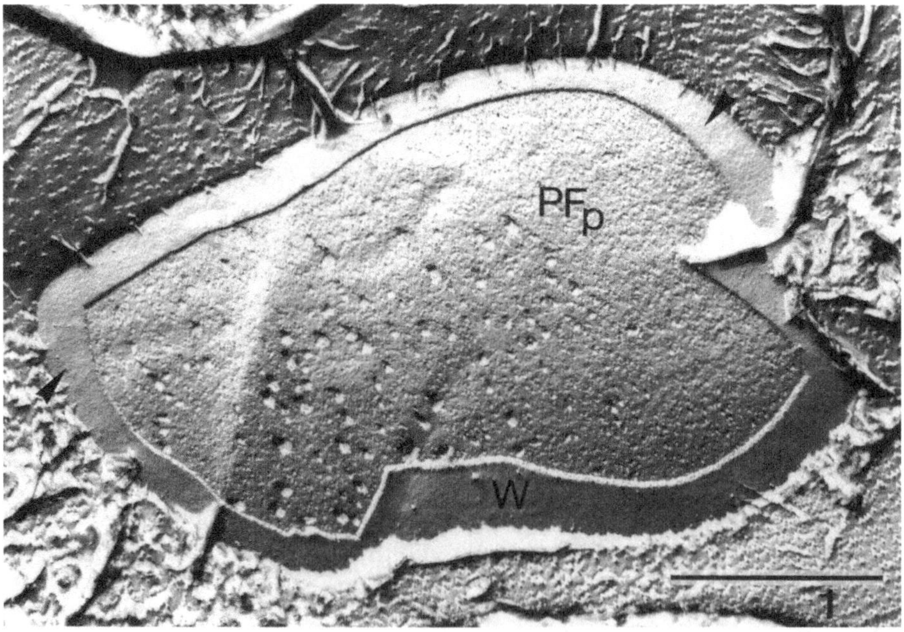

Fig. 2.42. Micrograph of a freeze-etch replica of *Chlamydomonas.* Nomenclature as for Figs. 2.39 and 2.41. As a result of the etching the fine structure of the cell wall *(arrowheads)* is now visible. Magnification bar in μm

vided that the partial pressure of the water vapour in the recipient is less than that of the ice. The sublimation rate (rate of ice recession) under constant vacuum conditions is mainly dependent upon the specimen temperature. Etching is most controllable at a temperature of $-100°$ C where up to 90 nm ice per minute can be removed. As a rule of thumb, it can be said that with a constant etch period the etch depth can be raised or lowered by a factor of 10 when the specimen temperature is raised or lowered by 10° C.

Since at $-100°$ C the partial and saturated water vapour pressures are similar, the possibility of recondensation of sublimed ice is reduced to a minimum. At lower temperatures, however, the saturated vapour pressure decreases considerably, making a contamination of the exposed surfaces unavoidable. This can only be combatted effectively at a very high vacuum (10^{-9} torr). Even at $-100°$ C one has to take precautions against recondensing water. This is done by positioning the knife assembly about 1-2 mm above the fractured object. Because of its much lower temperature ($-196°$ C), the knife acts as a cold trap for the sublimed water vapour. This procedure can be undertaken for both normal as well as double-replica specimen holders.

Etching is only meaningful in practice when the object is frozen in water, since cryoprotectants and high solute concentrations considerably alter the vapour pressure of the system. However, resuspending in water before freezing can result in the loss of peripherally associated membrane proteins. To avoid this, objects may be treated with glutaraldehyde (0.5-1%, v/v; 15-30 min) prior to resuspending in water. Optimal etching conditions must be determined for each new object, but periods longer than 20 min are seldom necessary.

Literature

Gross A (1979) Advances in ultrahigh vacuum freeze-fracturing at very low specimen temperature. In: Rash JE, Hudson CS (eds) Freeze-facture: methods, artifacts, and interpretations. Raven, New York, pp 127-139

Pinto da Silva P, Branton D (1970) Membrane splitting in freeze-etching. Covalently bound ferritin as a membrane marker. J Cell Biol 45: 598-605

2.6.5 Shadowing and Replica Formation

The principles of heavy metal shadowing which were dealt with under Section 2.3.1.3 also apply to production of freeze-etch replicas. However, the thickness of such Pt/C films (2 nm) is insufficient to withstand the subsequent cleaning steps. Thus an additional carbon layer (20-30 mm thick), which does not constitute a barrier to the electron beam in the EM, is deposited vertically onto the initial replica. With the appropriate specimen table it is now possible to do rotary shadowing as well as conventional unidirectional shadowing.

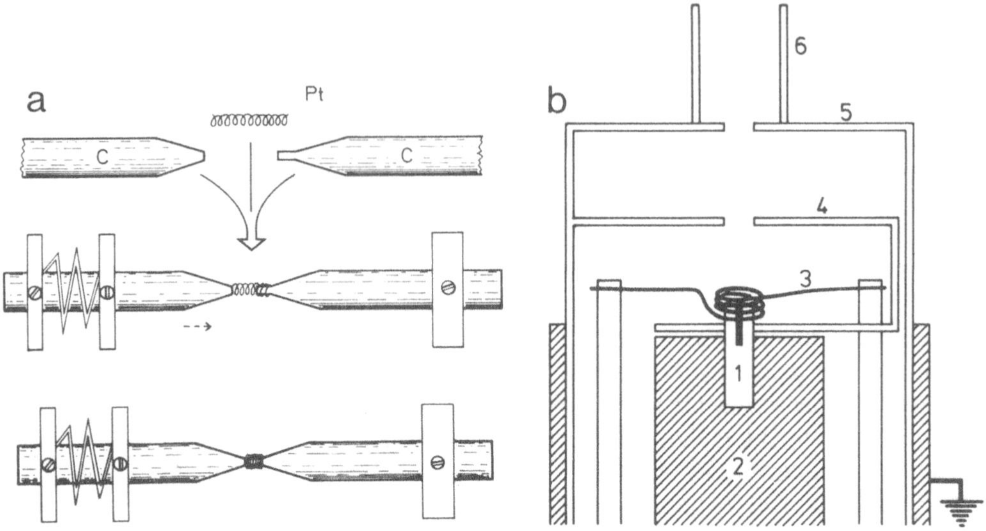

Fig. 2.43 a, b. Evaporators in freeze-etch machines. **a** Resistance heating evaporator (*Pt* platinum wire; *c* carbon or graphite rod). **b** electron beam evaporator. *1* Anode; *2* anode holder; *3* heating coil (cathode); *4* Wehnelt aperture; *5* electron brake; *6* deflection plate

2.6.5.1 Resistance-Heating Evaporation

Older freeze-etch machines are equipped with resistance evaporators for both the Pt/c and C-backing films. In each case the electrodes are pencil-like graphite rods (0.65 cm diameter), one of which is dampened and the other kept under spring tension. This enables a constant pressure to be maintained at the region of contact of the two electrodes during the evaporation. One electrode in each electrode pair has a flattened tip (2 mm in diameter). For the Pt/C evaporator the other electrode is so sharp that a small cylinder (1 mm in diameter) extends out from the tip (Fig. 2.43 a). A coil of platinum wire (0.1 mm in diameter, 6 cm long; previously wound tightly around a 1 mm-diameter metal jig) is pushed onto the cylindrical carbon tip. The electrodes are then brought into contact with one another and the platinum coil carefully pushed into position above the junction between the two electrodes. Evaporations are carried out with high current values (50-70 amp) but a low voltages (2 V for the Pt/C evaporator; 8 V for the C/C evaporator). The evaporation times are usually 5 s for the Pt/C electrodes and 10 s for the C/C electrodes. The heat which is developed at the junction of the electrodes is sufficient to evaporate both platinum and carbon, which then leave the evaporator through a small hole (0.5 cm diameter) in the metal shield which surrounds the electrodes.

2.6.5.2 *Electron Beam Evaporation*

Electron beam evaporators are present in nearly all modern freeze-etch machines. They are more expensive than the resistance heating evaporation, but are much more reliable, and several evaporations (15 for Pt/C; 5 for C/C) can be carried out without recharging the electrodes.

In this method, as already mentioned (see Sect. 2.3.1.3), electrons from a tungsten cathode fall onto an anode which contains the substance to be evaporated. The electron beam becomes accelerated due to the high voltage, and considerable heat is developed when it impinges upon the anode. The resulting vapourized carbon or platinum/carbon mixture then passes through a Wehnelt aperture before being directed onto the specimen via one or two deflection plates (see Fig. 2.43 b).

The anode in the Pt/C evaporator is a carbon rod into which a spigot of platinum is inserted. Providing it is not disturbed, the Pt/C evaporator does not have to be realigned and recentred for each new evaporation. This is often unavoidable for the C/C evaporator since, in contrast, much larger amounts of the anode are lost with each evaporation. With both evaporators the optimal evaporation period lies between 4 and 8 s.

2.6.5.3 *Measurement of Replica Thickness*

Before the vibrating quartz film monitor was introduced, it was customary to judge the thickness and quality of the replicas by the appearance of a piece of white card placed around the specimen table. Good evaporations gave rise to a brownish, silver-grey deposit on the card.

Modern freeze-etch machines are equipped with a thin-film monitor. This consists of a small quartz disc which is set into vibration electrically. The disc is mounted in a special holder which is adjacent to the specimen table. Evaporated carbon-metal mixtures which become deposited onto the quartz disc cause a change in the frequency of vibration. Frequency changes lying between 150 Hz and 200 Hz correspond to an optimal replica thickness.

Literature

Margaritis LH, Elgstaeter A, Branton D (1977) Rotary replication for freeze-etching. J Cell Biol 72: 47-56

Moor H (1971) Recent progress in freeze-etching technique. Philos Trans R Soc London Ser B 261: 121-131

2.6.6 Cleaning the Replica

After replication of the fractured specimen has been completed, the vacuum recipient is ventilated. The specimen supports, together with object and replica, are allowed to warm up and are then transferred with fine tweezers to a small Petri dish containing water. The replica is removed from the surface of the object by carefully inserting the specimen support at an angle into the water.

Replicas of animal material are relatively easy to clean: 2–4 h in 20–40% (w/v) chromic acid are usually sufficient. Those from plant sources present more difficulties due to the presence of a cellulosic cell wall. As a result these replicas are floated on 70% H_2SO_4 for at least 12 h. This concentration is reached stepwise over several hours; similarly, the replica must be brought stepwise back into water. The replicas are than transferred to 20% chromic acid for 1–2 h before returning them again to water. After being washed three times in dist. H_2O, the replicas are ready to be picked up (from below) onto EM grids. Coated grids are to be preferred for this purpose, since replicas on naked grids can easily be damaged when the grid is dried with filter paper.

A platinum loop is usually used for transferring the replica from one cleaning solution to another. Cleaning, i.e. the removal of contaminating biological debris, is carried out in small Petri dishes or watch glasses. If a replica sinks, rather than remaining afloat, and begins to curl up, one can usually reverse this effect by transferring the replica to 100% acetone or ethanol and then back into water.

Complementary replicas prepared with index grids as described above (Sect. 2.6.3.2) usually remain attached to the grids. The Vaseline between grid and replica can be removed by floating on chloroform for 20 min. With luck, grid and replica ramain together and can be cleaned as above, provided the grids are made of a resistant metal such as gold or nickel. If a lecithin-coated index grid has been used, the replica tends not to remain attached to the grid. In this case the replica should be picked up on a single-slot grid in order to be able to recognize the index numbers/letters which are visible in the replica.

2.6.7 Artifacts in Freeze Etching

There are many artifacts which can occur in the cource of the preparation of a freeze-etch replica. These are associated with each of the various steps in this technique and can be summarized as follows:

1. cryoprotectant-induced artifacts,
2. freezing artifacts,
3. fracturing artifacts,
4. etching artifacts,
5. shadowing or evaporation artifacts,
6. artifacts caused by cleaning the replica.

There are simply too many artifacts to be able to describe each one fully in a book of this nature. Instead, we refer the reader to the numerous excellent articles on this aspect of the freeze-etching technique. With the help of the various micrographs published in these articles one can easily "spot" the particular artifact in question.

Literature

Böhler S (1975) Artifacts and specimen preparation faults in freeze-etch technology. Balzers AG Publ FL-9496 DN6573. Balzers, Liechtenstein

Böhler S (1979) Artifacts and defects of preparation in freeze-etch technique. In: Rash JE, Hudson CS (eds) Freeze fracture: methods, artifacts, and interpretations. Raven, New York, pp 19-29

Miller KR (1979) Artifacts associated with the deep-etching technique. In: Rash JE, Hudson CS (eds) Freeze fracture: methods, artifacts and interpretations. Raven, New York, pp 31-36

Staehelin LA, Bertand WS (1971) Temperature and contamination dependent freeze-etch images of frozen water and glycerol solutions. J Ultrastruct Res 37: 146-148

Willison JHM, Brown RM (1979) Pretreatment artifacts in plant cells. In: Rash JE, Hudson CS (eds) Freeze fracture methods, artifacts, and interpretations. Raven, New York, pp 51-57

2.6.8 Using a Freeze-Etch Machine: a Practical Description

The great majority of freeze-etch machines are products of Balzers. Because of this we give below a short, step-by-step description of a typical run performed on a BA 360M freeze etcher equipped with EVM 952 electron beam evaporators and the Q8G 201D thin-film monitor. A conventional specimen table is present which can accommodate three specimens. Fracturing is accomplished with the help of a knife assembly.

A. Switching on

- Push valve control lever to position 0
- Press button ON/OFF (on)
- Open valve to compressed air cylinder, adjust pressure to 3 bar
- Switch on rotary pump
- Push valve control lever to position 2; vacuum gauge to position 2
- when the vacuum decreases below 0.01 bar, switch on the diffusion pump
- increase water supply in the event that the alarm bell rings
- leave to pump for at least 30 min

B. Cooling the specimen table

- Push valve control lever to position 1; vacuum gauge to position 1
- When the pressure is less than 0.01 torr, bring valve control lever to position 3 and vacuum gauge to position 3
- When pressure is less than 10^{-5} torr proceed further

- Connect liquid N_2 Dewar to the freeze etcher
- Open valve in liquid N_2 pipeline to specimen table (press TEMP/ADJUST button and set TEMP/ADJUST knob to 0.0)
- Open valve between N_2 gas cylinder and liquid N_2 Dewar
- Open valve on liquid N_2 Dewar; adjust gas pressure of N_2 from cylinder to 1.5 bar
- Press button temperature scale $-150°/50°$
- Wait until $-150°$ has been reached

C. Transfer of object to specimen table

- Place liquid Freon and small liquid N_2 Dewar with the frozen specimens near the vacuum recipient
- Push valve control lever to position 0
- Turn air-lock switch to VENT; raise the recipient. Turn air-lock switch to 0
- Push valve control lever to position 2; swing out C/C evaporator; bring knife assembly in posterior position; point specimen table with liquid Freon; transfer frozen objects with tweezers to specimen table; fix in position; remove protective cap from vibrating quartz; swing back C/C evaporator, lower recipient
- Push valve control lever to position 1; vacuum gauge in position 1
- When the pressure is less than 0.01 torr, push valve control lever into position 3; vacuum gauge to position 3
- Bring specimen table temperature to $-100°$ C or $-110°$ C (TEMP/ADJUST knob to 2.9-3.2; Temp. scale $-80°-120°$)
- cool knife assembly to $-192°$ (press COOLING button)
- Wait 15 min (pressure in recipient should then be less than 10^{-5} torr)

D. Fracturing the specimen

- Press button LAMP
- Raise knife to the highest position with left-hand wheel control (anticlockwise)
- Move right-hand wheel control anticlockwise through 360° (cutting movement)
- Move left and right hand wheel movements until the specimen is fractured with an even, plateau-like, surface

E. Etching

- After the last cut retract knife through 120 µm (three anticlockwise turns with left-hand wheel control)
- Position knife immediately above the specimen
- wait for the desired etching period; then move knife away

F. Replication

- Press button LAMP (switch off)
- Press button POWER on thin film monitor [Pt/C evaporation]
- Switch on FILAMENT 1
- Turn POWER HIGH, VOLTAGE ON knob to maximal position
- Adjust HIGH VOLTAGE to 1800 V
- Adjust filament current to 70 mA
- Select RANGE of 0.3
- Monitor frequency change (zero point is calibrated to 50 Hz; frequency values between 150-200 Hz are usual)
- Switch off FILAMENT, HIGH VOLTAGE

- Return FREQUENCY to zero C/C evaporation
- Switch on FILAMENT 2
- POWER HIGH VOLTAGE ON to maximum
- HIGH VOLTAGE to 2400 V
- FILAMENT to 90 mA
- RANGE 0.3
- Frequency change of 180–200 Hz
- Switch off FILAMENT; HIGH VOLTAGE
- Return FREQUENCY to zero

G. Removing the replicas from the recipient

- Push valve control lever into position 0
- Air lock switch to VENT; raise recipient. Turn air lock switch to 0
- Push valve control lever to position 2
- Free specimen supports from specimen table
- Remove specimen supports plus replica(s) from recipient

H. Preparing the freeze-etch machine for the next run

- Press button COOLING (switch off)
- Close valves on liquid N_2 Dewar and N_2 gas cylinder
- TEMP/ADJUST knob to 10
- Temperature scale to $-150/+50$
- Replace cap onto vibrating quartz
- Press HEATING button (switch on)
- When the specimen table temperature reaches $+10°$ C, turn TEMP/ADJUST knob to "0.0"
- Press button COOLING (Valve opens but warm N_2 gas flows through the pipeline)
- Thaw knife assembly externally with a hair dryer
- If necessary change razor blade (or bring new cutting edge into position)
- Realign Pt/C and C/C evaporators
- Press HEATING button (switch off)
- Lower recipient

I. Evacuating the recipient

- Push valve control lever into position 1; vacuum gauge into position 1
- When the pressure is less than 0.01 torr, shove valve control lever into position 3; vacuum gauge to position 3.

J. Switching off

- Push valve control lever to position 2
- Switch off diffusion pump
- Allow 30 min to cool (leave water running)
- Push valve control lever to position 0
- Switch off rotation pump
- Close valve to compressed air cylinder
- Turn water off
- Press button ON/OFF (off)

3 Methods for SEM

The surface of an object which is to be investigated by scanning electron microscopy (SEM) must have the following characteristics (Reimer and Pfefferkorn 1977):

- it must be free from foreign particles (e.g. dust, glandular exudates, etc.),
- it must be vacuum stable,
- it must remain stable after exposure to the electron beam,
- it must emit a sufficient number of secondary electrons (see Chap. 1.1.2) and should develop as few surface charges as possible.

Some biological structures fulfill automatically these prerequisites, e.g. the hard exoskeleton of insects or crustaceans, various mineralized structures (teeth, bone, diatom frustules), as well as many structures of plant origin (pollen, wood etc.) In the majority of cases, however, biological objects have to be treated in order that the above requirements may be met. The preparation will differ according to the nature of the object and the method of investigation (imaging with secondary and/or back-scattered electrons, element analysis). Clearly in a book of this nature it is not possible to describe all of the various possibilities which exist; the reader is therefore referred to the more detailed accounts which are given at the end of this introduction.

If the object has already been investigated for TEM purposes and reliable methods on isolation, fixation etc. are at hand, the same method should also be employed for SEM. Should such data not be available, it is always advisable to try to obtain information on the structure and morphology of the object in question using a number of different techniques in order to be certain of the veracity of the SEM results obtained.

Despite having carried out the appropriate preparation procedure, it sometimes happens that the surface is contaminated with large particles or that the desired surface is still occluded within the tissue. This problem can be overcome in some cases with the help of a built-in micromanipulator (Pawley et al. 1975).

The SEM produces images of surfaces which have a strong three-dimensional plasticity. Because of this there is a tendency to interpret the image in real, three-dimensional, terms. One should, however, be aware that the SEM image differs from the true-life object in a number of ways, both as a result of the method of preparation as well as the way in which the image is constructed. Because of this the production of stereo pairs is an absolute requirement for a detailed three-dimensional analysis of an object. Nevertheless, one has to realize that the recognition of artifacts in SEM is much more difficult than is the case with TEM images.

Literature

Chatfield EJ (1978) Introduction to stereo scanning electron microscopy. In: Hayat MA (ed) Principles and techniques of scanning electron microscopy, vol VI. Biological applications. Van Nostrand-Reinhold New York, pp 47-88

Falk RH (1980) Preparation of plant tissues for SEM. Scanning Electron Microsc 1980/II: 79-88

Goldstein JJ, Newbury DE, Echlin P, Joy DC, Fiori C, Lifshin E (1981) Scanning electron microscopy and X-ray microanalysis. Plenum, New York London, pp 1-673

Haggis GH (1982) Contribution of scanning electron microscopy to viewing internal cell structure. Scanning Electron Microsc 1982/II: 751-763

Maugel TK, Donar DB, Creegan WJ, Small EB (1980) Specimen preparation techniques for aquatic organisms. Scanning Electron Microsc 1980/II: 57-78

Murphy JA (1982) Considerations, materials, and procedures for specimen mounting prior to scanning electron microscopic examination. Scanning Electron Microsc 1982/II: 657-696

Newbury DE, Joy DC, Echlin P, Fiori CE, Goldstein JI (1986) Advanced scanning electron microscopy and X-ray microanalysis. Plenum, New York London, pp 1-454

Nowell JA, Pawley JLB (1980) Preparation of experimental animal tissue for SEM. Scanning Electron Microsc 1980/II: 1-20

Pawley JB, Hayes TL, Nowell JA (1975) Microdissection. In: Hayat MA (ed) Principles and techniques of scanning electron microscopy. Biological applications, vol III Van Nostrand-Reinhold, New York, pp 17-34

Postek M, Howard K, Johnson A, McMichael K (1980) Specimen preparation. In: Scanning electron microscopy: a students handbook. Ladd, Burlington, VT, pp 115-240

Reimer L, Pfefferkorn G (1977) Raster-Elektronenmikroskopie, 2nd edn. Springer, Berlin Heidelberg New York, pp 1-282

Revel J-P, Barnard T, Haggis GH (1984) The science of biological specimen preparation for microscopy and microanalysis. Scanning Electron Microscopy Inc, AMF O Hare, II, pp 1-245

Rosenbauer KA, Kegel BH (1978) Rasterelektronenmikroskopische Technik. Präparationsverfahren in Medizin und Biologie. Thieme, Stuttgart, pp 1-241

Watson LP, McKee AE, Merrell BR (1980) Preparation of microbiological specimens for scanning electron microscopy. Scanning Electron Microsc 1980/II: 45-56

3.1 Conventional Methods of Preparation

3.1.1 Introduction

In the following sections we will present a number of preparatory procedures which are widely used in medicine and biology. These methods have now become standard techniques. Usually the preparation of an object for SEM involves the following steps:

a) selection, removal and cleaning up the desired tissue, portion or organ;
b) stabilization of the objct (fixation). This is done either by immersion (smaller objects) or by perfusion (larger objects);
c) washing and dehydration (in the case of chemically fixed objects). Post-fixation washing serves not only to remove excess fixative but also to clean up the object once again. This latter process also takes place during the subsequent dehydration in an organic solvent;
d) drying and mounting the object. This is usually carried out with the help of a critical point-drying apparatus using appropriate SEM object holders (mainly of aluminium);
e) Creating or increasing surface conductivity. This step is achieved either in an evaporation unit or in a sputtering apparatus (see Sects. 3.1.5 and 3.1.7). Depending on the type of fixation (whether or not OsO_4 has been used, which in itself imposes a degree of conductivity to the specimen) this step may not be necessary.

Objects prepared in this way can normally be inserted into the SEM straight away. If stored, they must be maintained in a dry, dust-free environment.

One must not forget that the various preparatory steps given above can lead to changes in the surface of the object to be studied. The extent to which this can occur depends on the type of cell or tissue involved and sometimes on their metabolic status at the time of fixation (Boyde et al. 1977). This can necessitate modifications in one or more of the standard steps.

It is imperative that the object should always be immersed in liquid during the various preparatory steps right up to the drying stage. Sensitive surfaces can easily be damaged if they become exposed momentarily to air (e.g. the cilia of epithelia stick together when brought out of liquid). It is also to be recommended that liquids be changed frequently during precleaning, fixation and washing. Moreover it is better when changing solutions that this be done in several steps, i.e. by exchanging only a portion of the liquid each time. Transferring the object from one vessel to another should, if possible, be avoided. If this must be done one should always take care that the object remains covered with liquid; this requirement is easily fulfilled by using a pipette to transfer the object or transferring it in a drop of liquid hanging from a needle.

In general, the cleanliness, both of glassware and solutions, and the safety precautions which are necessary for TEM are also to be upheld in the case of SEM preparations. Particulate impurities in solutions or on the walls of glass vessels can lead to artefactual deposits on the surfaces of the objects to be investigated. These can often only be detected in the SEM itself.

Literature

Boyde A, Bailey E, Jones SJ, Tamarin A (1977) Dimensional changes during specimen preparation for scanning electron microscopy. Scanning Electron Microsc 1977/I: 507-518

3.1.2 Specimen Size; Handling Specimens and Exposing Surfaces

The type of SEM available often determines the size of specimen which can be investigated. Older SEMs have a standard object holder corresponding to a maximal translation of 15 mm in both x and y directions. Special macro object holders are, however, available, which enable a translation of up to 100 mm. By contrast, objects as large as a child's fist can be mounted on the standard object holder of modern SEMs which have a translation capacity of 200/200 mm.

Whether or not a specimen is to be chemically fixed also plays a role in determining its maximal size for SEM. Thus objects fixed by the immersion procedure should not exceed 1 mm^3 (see Chap. 1.2.2). This is also the size limit for cryofixed objects (see Sect. 3.1.3.2) whereby the chance of ice crystal damage increases considerably at larger volumes. Specimen size is also important for objects drying out of organic solvents; larger objects tend to develop tears or cracks more frequently than smaller ones. Thus, although larger objects can easily be observed in the SEM from the optical point of view, it is always better to choose the smallest possible individuals when studying external surfaces in order to reduce the possibility of preparation artifacts.

When internal surfaces are to be investigated, the tissue or organ has to be first reduced in size. In so doing structural deformations can easily occur when cutting implements such as scissors or a scalpel are used for this purpose. Much safer are razor blades, whose cutting edge should not be used more than once. In addition it is recommended that this process be carried out on objects which have already been stabilized with an aldehyde fixative (either by perfusion or immersion). In contrast, the cut surfaces of native, unfixed material tend to reveal much more damage (Boyde 1976). Cutting is best performed when the object is pinned down onto a soft base, e.g. dental wax or paraffin, and is most easily executed when the object is immersed in 50–70% ethanol, which makes the object a little bit more leathery and easier to cut.

Isolated epithelia or thin pieces of tissue have the tendency to roll up during their preparation, particularly during dehydration and drying. In order to prevent this, such objects should be pinned down right from the start and remain so until the end of the drying phase.

Very small objects, e.g. red blood cells, can be stuck down onto glass surfaces (cover slips, pieces of microscope slides) covered with albumin or gelatine, either before or after an immersion fixation.

3.1.2.1 Cleaning Surfaces

Only seldom is the surface which is to be investigated free from contamination. The substances responsible for this range from general dirt in the case of aquatic organisms and some terrestrial organisms to complex organic secretions (slime, blood etc.).

Although larger particles can often be removed with the help of a fine needle or brush (using a stereomicroscope), cleaning in the case of aquatic organisms or tissues/organs is best achieved by rinsing in filtered fresh- or seawater or with isotonic physiological solutions, respectively. These solutions can also be used as wash solutions after fixing. Unfortunately, there are disadvantages with some physiological solutions: the frequently used NaCl-based type occasionally leads to structural changes due to the high Cl^- concentration, and Krebs-Ringer solution is known to cause cells to swell. This means that rinsing with these agents should be performed quickly and with several changes of solution. When necessary, specimens can be additionally cleaned after fixing by using either isotonic 1-10% (v/v) HCl to remove secretions, or phosphate or cacodylate buffers (pH 7.2) which help get rid of serum proteins. For more details on rinsing agents the reader is referred to Rosenbauer and Kegel (1978).

Small aquatic microorganisms and cell suspensions can be separated from their liquid environment by centrifugation. They can also be cleaned by repeated resuspension and centrifugation in an appropriate physiological solution. This cleaning process is, of course, automatically continued during fixation and dehydration.

Mechanical methods of cleaning, e.g. ultrasonics or spray guns, have not enjoyed much success. Only for very hard structures, e.g. teeth and boone, are such methods to be recommended. With softer material, e.g. tissue, epithelia or even single cells, considerable damage can be inflicted. Etching is also a procedure that can essentially only be carried out on hard materials (e.g. see Fig. 3.1a). In this case the surface of the tooth is polished with a suitable abrasive and then sonicated in acetone. After drying in air the tooth is etched with 5-10% (v/v) HCl or H_3PO_4 for 1 min ("Ionetching" is also possible with hard materials; the results, however, are difficult to interpret). The etched tooth is rinsed in distilled H_2O, dehydrated in ethanol or acetone and coated after drying.

A relatively simple procedure for the effective portrayal of calcified structures (see for example Fig. 3.1b) consists of treating the bone to high temperature (105° - 120° C) for 1 h. After cooling to room temperature, the bone is split open by placing on a hard surface and hitting it with a hammer. The bone dust is blown off with a gas pistol, and the bone pieces mounted onto a SEM object support and the fractured surface coated. Unfortunately this method does not produce reproducible fracture planes, nor can it be said to represent the natural state of the bone.

Literature

Boyde A (1976) Do's and dont's in biological specimen preparation for the SEM. Proc Annu Scan Electron Microsc 9/1: 683-690

Rosenbauer KA, Kegel BH (1978) Rasterelektronenmikroskopische Technik. Präparationsverfahren in Medizin und Biologie. Thieme, Stuttgart, pp 1-241

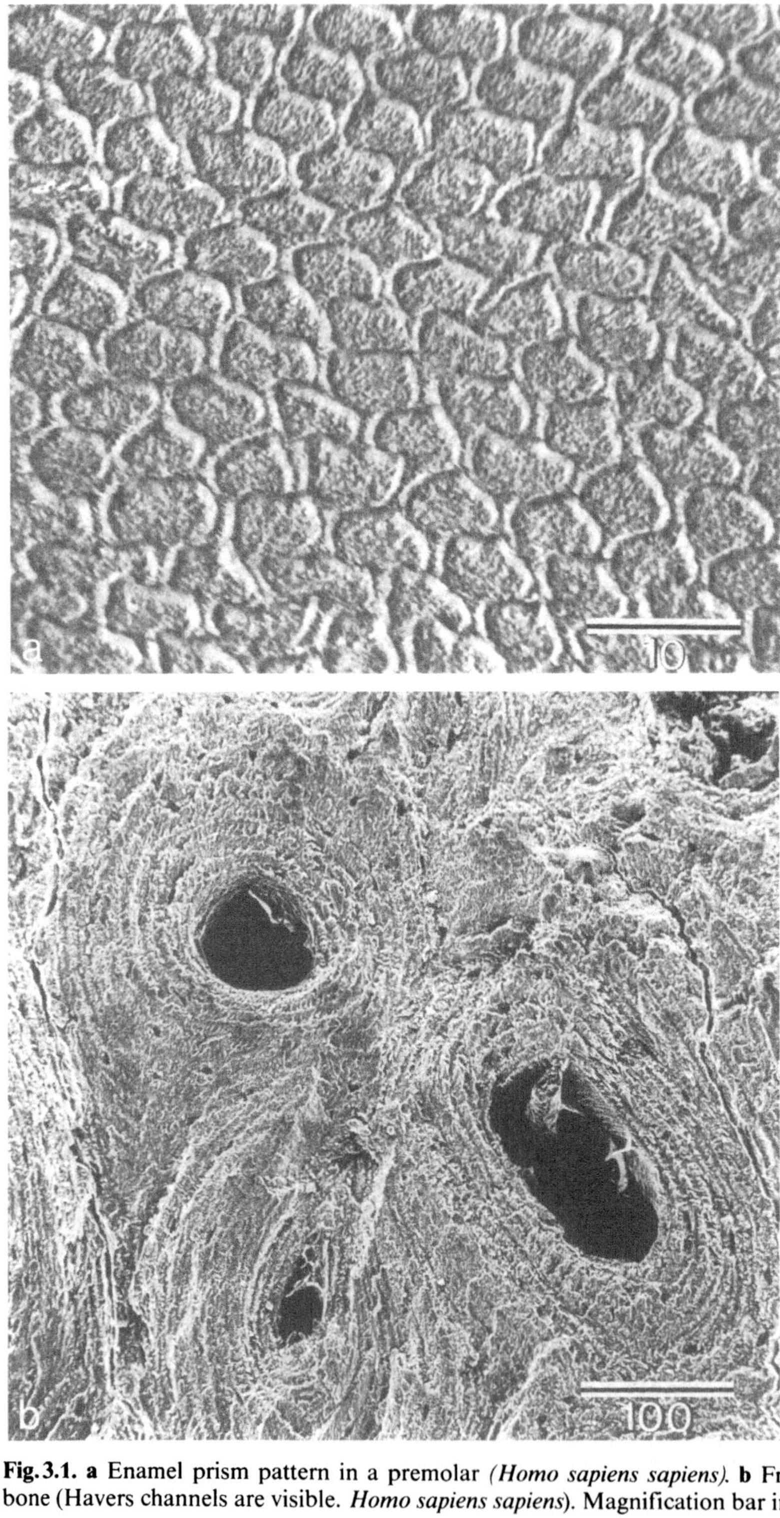

Fig. 3.1. a Enamel prism pattern in a premolar *(Homo sapiens sapiens).* **b** Fractured femur bone (Havers channels are visible. *Homo sapiens sapiens*). Magnification bar in μm

3.1.3 Stabilization

As a rule biological specimens must be dehydrated and dried before inserting into a SEM. Apart from hard objects such as teeth, bone and wood, all others must be stabilized to prevent structural changes from occurring. Usually this is done via a chemical fixation, but a physical fixation (freeze fixation) can also be employed.

3.1.3.1 Chemical Fixations

In general, the principles and methods already described under Chapter 2.1 for TEM are applicable. In order to prevent postmortal autolysis the fixatives have to penetrate as quickly as possible and be in a form or concentration so that isotonic conditions can be retained during the fixation.

Perfusion Fixation and Microinjection

Fixation by perfusion (see also Chap. 2.1.1.4) is recommended for all larger animals (both vertebrates and invertebrates). Most frequntly used is a buffered (pH 7.2-7.5; phosphate or cacodylate) glutaraldehyde (1-4%, v/v) solution, at a temperature corresponding to that of the animal in question (for details of preparing fixatives see Chaps. 2.1.1.2 and 2.1.1.3)

When perfusing, it is important that the blood or body fluid pressure of the animal is not neglected. One can perfuse either by forcing the blood out first before administering the fixation solution or by adding anticoagulants to the fixative and pumping blood and fixative out together (see also Chap. 2.1.1.4.1). The latter method has the advantage that it reduces the danger of autolytic processes. (Consult appropriate textbooks for dosages of anticoagulants).

If a perfusion fixation is not possible, the fixation solution can be injected directly into the tissue or organ to be studied. This is most appropriate for the vertebrate eye or the mixocoel of insects. Pressure differences should also be avoided. When injecting, the rule is: only as much fixative should be injected as there is body fluid leaving the organ or object.

The object is then well rinsed with buffer (phosphate, cacodylate) after the primary fixation. A secondary fixation with OsO_4 (1-2%, w/v; higher concentrations should be avoided since they can cause a loss of microfilaments leading to cell deformation) can then be performed either by perfusion or microinjection, or, after carefully opening up the object, by immersion. Depending on the size of the object (cutting the object into smaller pieces may be necessary) the duration of the secondary fixation can extend up to 24 h.

A double fixation with an aldehyde followed by OsO_4 is not only of general use, it is specifically recommended when fractured surfaces are to be prepared.

Objects fixed in this way also reveal fracture planes in the intracellular spaces, whereas objects fixed in OsO_4 alone tend to fracture primarily intercellularly.

Fixing with OsO_4 serves essentially two purposes. As already mentioned (see Chaps. 1.2.2 and 2.1.1) OsO_4 prevents the extraction of cell lipids during dehydration with organic solvents. Secondly OsO_4 increases considerably the conductivity of the specimen and thereby gives rise to a much higher emission of secondary electrons. It may even not be necessary to coat the surface with a heavy metal if sufficient osmium is present in the specimen. On the other hand, if a secondary fixation with OsO_4 is not carried out, the necessity to coat the object with a conducting metal (see Sect. 3.1.7) becomes even more mandatory.

Immersion Fixation

This type of fixation has already been dealt with (see Chap. 2.1.1.4.1) for TEM. All that is necessary here is to briefly summarize the salient features of the method and to point out specific fixation schedules which are in use for SEM.

The organisms to be fixed are usually small. If very agile, they should be subjected to low temperatures (4° C) before application of the fixative. The fixation solutions should be isotonic (osmolarity adjusted with NaCl or sugar) with the culture medium or body fluid and normally contain 1-4% (v/v) glutaraldehyde in cacodylate or phosphate buffers. The duration of the fixation can vary but 2 h at room temperature is usually sufficient for objects not larger than 1 mm^3. Longer primary fixations with aldehydes often lead to holes and tears in the plasma membrane (Boyde and Maconnachie, 1981). Aldehyde fixations tend to induce a light swelling of the cells which, of course, can cause changes in the surface structure of the object. Such volume changes are, however, usually compensated by the tendency of the object to shrink during the later drying phase (see Sect. 3.1.5).

After the primary fixation the objects are washed in the isotonic fixation vehicle. Objects for SEM can be stored up to several months at 4° C in such a medium. However, because of the tendency for bacteria to grow in it, phosphate buffer should not be used for such long periods of storage. Cacodylate is better and should be occasionally exchanged for fresh buffer.

As explained in the previous section on perfusion fixation, a second fixation with OsO_4 is always to be recommended. In order to increase the relative amounts of osmium deposited in the object, mordanting agents such as tannic acid and thiocarbohydrazide (see also Chap. 2.1.1.3.4) are often included in osmium fixations for SEM.

Tannic acid is usually given as a 1-7% (w/v) buffered solution, whereas thiocarbohydrazide is normally a saturated aqueous solution (in both cases the solutions should be filtered before use). The specimens are immersed in these solutions at room temperature after the OsO_4 fixation (usually 1%, w/v; followed by several rinses in buffer or water). After washing in water (several changes over 20 to 30 min) the specimens can be immersed again in a buffered OsO_4 solution. This procedure can be repeated several times, hence the terms OTO and OTOTO for this method. If the objects are small, and one is primarily interested in the external surface, only brief immersions (10-20 min) are necessary. For larger objects, and when internal surfaces are to be investigated, longer (1-2 h) periods should be used (see, for example, Murakami et al. 1983). Often objects fixed according to the OTOTO method do not require a metal coating before insertion into the SEM. On the other hand, one must be aware of the fact that thick (several nm) osmium deposits on the cell/tissue surface may be formed (probably due to an interaction with components of the glycocalyx) which therefore can falsify the situation somewhat.

The OsO_4-sublimate fixation of Parducz (1967) has proved particularly useful for the preservation of cilia structure and for protozoans in general. The fixation solution is made up from 2% (w/v) OsO_4 and saturated $HgCl_2$ mixed together immediately before use in the proportion 6:1 (v/v). Four parts of the fixation solution are usually added to 1 part of the culture medium; the duration of the fixation is short (15 min) and is performed at room temperature. Afterwards the specimen is rinsed several times in destilled water.

Fixation in Osmium Fumes

This method can be applied only to those objects which are already dry. Specimens are brought together in a sealed glass container with either crystalline OsO_4 or a solution of the same, making sure that the osmium does not come into direct contact with the object. This method is useful for bacteria which have been growing on agar (Rosenbauer and Kegel 1978).

Literature

Boyde A, Maconnachie E (1981) Morphological correlations with dimensional changes during SEM specimen preparation. Scanning Electron Microsc 1981/IV: 27-34

Kelley RO, Dekker RAF, Bluemink JG (1975) Thiocarbohydrazide-mediated osmium binding: a technique for protecting soft biological specimens in the scanning electron microscope. In: Hayat MA (ed) Principles and techniques of scanning electron microscopy, vol IV. Biological applications. Van Nostrand-Reinhold, New York, pp 34-44

Murakami T, Lida N, Taguchi T, Oktani O, Kikuta A, Ohtsuka A, Itoshima T (1983) Conductive staining of biological specimens for scanning electron microscopy with special reference to ligand-mediated osmium impregnation. Scanning Electron Microsc 1983: 235-246

Murphy JA (1980) Non-coating techniques to render biological specimens conductive / 1980 update. Scanning Electron Microsc 1980/1: 209-220

Parducz B (1967) Ciliary movement and coordination in ciliates. Int Rev Cytol 21: 91-128

Rosenbauer KA, Kegel BH (1978) Rasterelektronenmikroskopische Technik. Präparationsverfahren in Medizin und Biologie. Thieme, Stuttgart, pp 1-241

3.1.3.2 Cryofixation

In some cases, especially when an element analysis is envisaged, a chemical fixation is disadvantageous (Panessa-Warren 1983, Zierold 1983). Here a freeze fixation should be carried out. The frozen object can then be brought directly onto a cold stage in the SEM and viewed after partial sublimation of ice, or investigated after freeze drying (see Sect. 3.1.5.2)

The principles of freezing biological specimens dealt with under Chapter 2.6.2 also apply here: one freezes as quickly as possible in order to prevent ice-crystal formation. Freezing is accomplished in a liquified cryogen, e.g. Freon 12 or 22, or propane. One should, however, note that only the outermost surface layers (several μm thick) of large objects can be adequately preserved in this way. Moreover, aquatic organisms frozen together with culture medium often show surface damage. In general, freeze-fixed objects reveal a poorer structural preservation of surface structures than is the case with objects fixed chemically.

Literature

Aldrian AF (1980) Studies on frozen specimen in the SEM. Micron 11: 261-265

Panessa-Warren BJ (1983) Basic biological X-ray microanalysis. Scanning Electron Microsc 1983/II: 713-723

Zierold K (1983) X-ray microanalysis of frozen-hydrated specimens. Scanning Electron Microsc 1983/II: 809-826

3.1.4 Dehydration

Chemically fixed objects must be dehydrated before drying. This cannot be achieved by simply allowing to dry in the air because of the considerable surface tension effects involved. Either the object is subjected to freeze drying (see Sect. 3.1.5.2) or it is dehydrated through substitution.

Once again the principle is the same as in TEM (see Chaps. 1.2.1 and 2.1.2); the object is immersed in 20-30% (v/v) ethanol after fixing and washing, and by transferring through more concentrated solvent solutions it is finally brought into absolute ethanol or acetone. The speed with which this can be done, and the number of individual steps involved, depends on the size of the object and its properties, but takes usually 10-60 min per step.

During the dehydration steps the specimens must on no account be allowed to dry out. With some objects lower solvent concentrations (20-40%) can lead to a swelling of the object, sometimes resulting in tears in the surface. This can be prevented to a great extent by the inclusion of divalent cations (concentration 1 mM-1 M) in the solvent solutions. Conversely higher solvent concentrations (70-90%) can lead to shrinkage. In general, acetone tends to induce greater swelling/shrinkage effects than ethanol.

An alternative to an acetone/ethanol dehydration series is a DMP (2,2 dimethoxy propane) dehydration. In this case the object is transferred to 100% DMP after fixing and washing. After 2 min the DMP solution is replenished with fresh solution and allowed to stand for up to 60 min. Although it is simpler to use, volume artifacts tend to occur more frequently with a DMP dehydration than with acetone/ethanol.

Finally, the object can be removed from the absolute solvent concentration and dried either directly, or over a number of intermediate substances (see below, Sect. 3.1.5).

3.1.5 Drying

After water has been removed from the object, the dehydrating medium (organic solvent) also has to be removed. This consitutes "drying" and has to be done very carefully in order to prevent deformation at the surface during the phase change from liquid to gas. Such artifacts can be successfully avoided by the critical point method (Sect. 3.1.5.1) and by freeze drying (Sect. 3.1.5.2). Of the two, the former is quicker and more preferable; moreover critical point drying apparatuses are generally much cheaper than those necessary for good quality freeze drying.

Only seldom is it possible to air-dry specimens successfully; in comparison to critical point drying it is always inferior (cf. Fig. 3.2a with 3.2b). Structural damage of this nature is, however, less extreme when the objects are infiltrated with a second solvent, e.g. ether, after dehydrating in acetone or ethanol.

3.1.5.1 Critical Point Drying

Principle: The object is transferred from an organic dehydration medium (acetone/ethanol) via an intermediate substance (freon 113 or amyl acetate) and then into the "drying" medium (usually CO_2 or a fluorinated hydrocarbon). Substitution of the drying medium for the intermediate medium then occurs in a chamber which is cooled and put under pressure (during this procedure the object dare not be allowed to dry out). When the intermediate substance has been completely removed the chamber is warmed up to that point (the critical point) where the density of the drying medium is the same in both the liquid and the gas phase. A phase boundary then exists no longer, and the drying medium can be released from the chamber; the object is then dry.

There are a number of different critical point apparatuses available commercially. Those with a large observation window at the side allow for a better optical control of liquid levels as compared to those with a window at the top of the chamber. On the other hand, it is not easy to judge the progress of substitution (turbulence, for example, is not easy to recognize) with a side window.

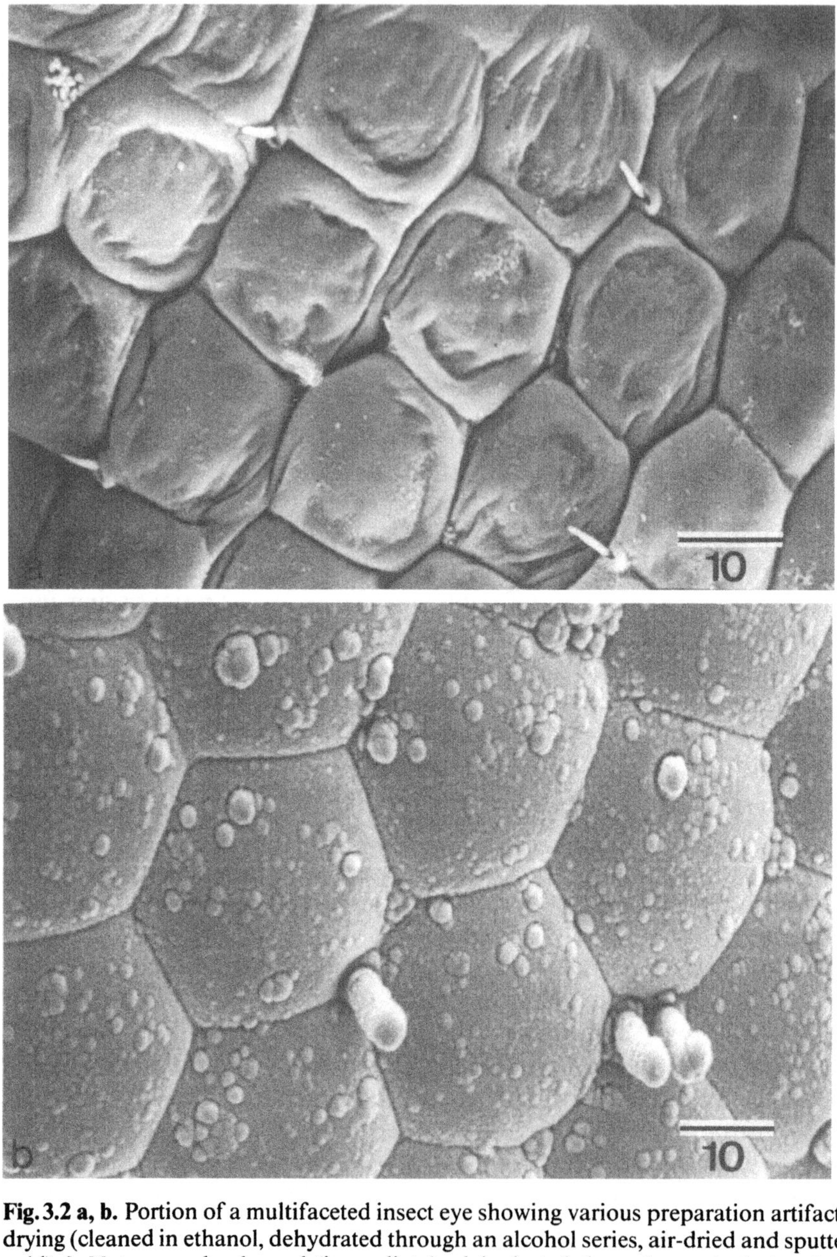

Fig. 3.2 a, b. Portion of a multifaceted insect eye showing various preparation artifacts. **a** Poor drying (cleaned in ethanol, dehydrated through an alcohol series, air-dried and sputtered with gold). **b** Not properly cleaned (immediately dehydrated through an alcohol series without cleaning, critical-point drying with CO_2 and sputtered with gold). Magnification bar in μm

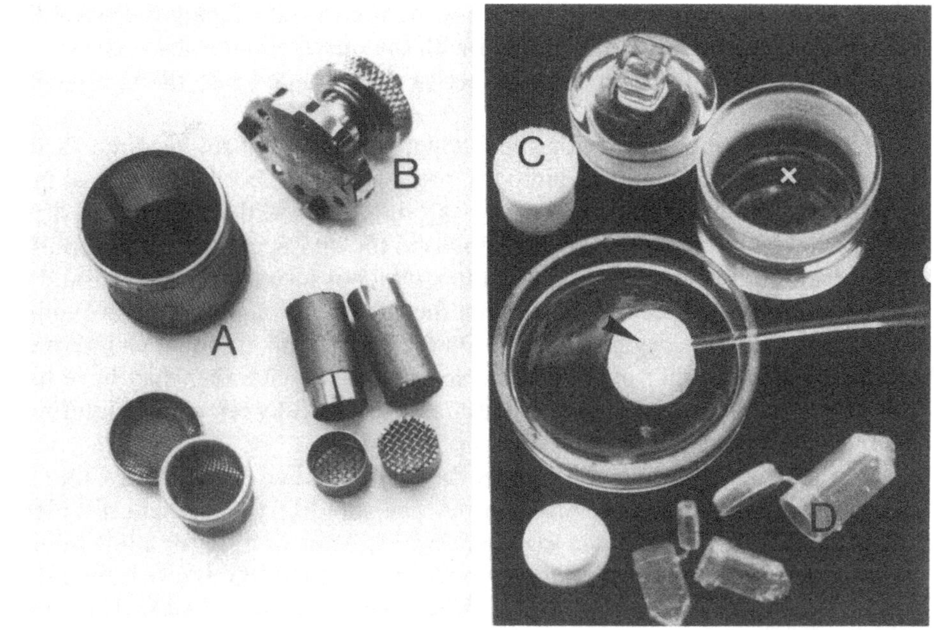

Fig. 3.3. Vessels/holders for critical point drying. *A* wire basket; *B* clamping device for grids; *C* plastic vial; *D* perforated BEEM-capsule. *White cross* indicates a weighing dish; *black arrowhead* shows a plastic vial into which a specimen is being pipetted

Clearly the best models are those by which both temperature and pressure in the chamber can be precisely controlled and registered.

Only large objects can be brought into the drying chamber directly. Smaller objects should first be placed in vessels which are then inserted into the chamber. Such vessels should allow a good exchange of liquids to occur but should also prevent the specimen from being washed out. Commercially available wire mesh containers (see Fig. 3.3) are usually only suitable for robust specimens; more sensitive objects should be placed in porous plastic containers (Fig. 3.3) such as those which act as chambers in automatic fixing (dehydration) and embedding machines. For objects with long and sensitive appendages it is best to use embedding (BEEM) capsules which have been perforated with a needle. For very small objects, e.g. isolated cells or epithelia, small "boats" made of aluminium foil can be used (they should be used only once to prevent contamination).

The most frequently used drying media are freon 13 or liquid CO_2. Freon 13 is much more expensive, but if correctly used, CO_2 can give just as good results. If Freon 13 is used, the intermediate substance is Freon 113; with liquid CO_2 the specimens are either transferred via isoamyl acetate or directly from 100% ethanol/acetone.

It is imperative that the specimens do not come into contact with the air when being transferred from the dehydrating medium into the intermediate medium. If porous plastic containers are to be used as the drying container they should be placed in a Petri dish filled with the intermediate substance (see Fig. 3.3); they rapidly fill up and the object can be pipetted into the intermediate medium.

The drying container with specimen(s) in intermediate medium is then brought as quickly as possible into the precooled chamber of the critical point apparatus. The chamber is sealed and carefully filled with the liquefied drying medium. When transferring the specimens to the drying chamber it is important that the amount of intermediate medium, which is also transferred, should not be too small, since there is a danger that the specimens might dry before coming into contact with the drying medium. On the other hand, if too much intermediate medium is brought over with the specimens the chamber may have to be rinsed several times with drying medium in order to fully remove the intermediate medium before the drying process is started.

The pre-rinsed drying chamber is then slowly filled with liquid CO_2. This mixes with the intermediate medium but one should try to prevent turbulence which might result in damage to the specimen. After waiting for a few minutes (with larger specimens even longer) a portion of the liquid mixture is then slowly released from the chamber and replaced with fresh liquid CO_2. This procedure is repeated several times until the intermediate medium is completely replaced with liquid CO_2. In general: the larger the specimen or the larger the volume of intermediate medium brought into the drying chamber, the greater the number of liquid CO_2 changes necessary.

Before the actual drying process is begun, the level of liquid CO_2 in the chamber is checked and regulated according to the maker's instructions for the particular apparatus. The chamber is then slowly warmed up until the phase boundary disappears (critical temperature for CO_2 is 31.3° C; for freon 13 it is 28.9° C); during this process one must again take care to avoid turbulence. The temperature is then allowed to increase a few ° C (with liquid CO_2 to 36-40° C; with Freon 13 to 33-36° C) in order to prevent a recondensation when the gas is released. If the pressure increases too much during the heating, one should carefully open the release valve to the drying chamber.

When the heating phase has been completed, the valve to the drying chamber is carefully opened (the high temperature should be maintained constant during the gas release). When the pressure in the chamber has been raised to that of atmospheric pressure the specimen can be removed from the chamber. If the specimen is not used immediately it must be stored in an exsiccator over an appropriate drying substance, e.g. silica gel, otherwise moisture from the air may condense upon it.

Critical point drying gives rise to excellent results for the great majority of biological objects. Nevertheless, artifacts can occur which are usually due to some mistake in the procedure. The major pitfalls are:

1. the object has become exposed to air during the transfer from dehydrating solvent into the intermediate medium or from the intermediate medium into drying medium,

2. the object was not subjected to a sufficient number of rinses in the liquefied drying medium before the drying process was started,
3. turbulence has occurred either during the mixing of one medium with another or as a result of heating too quickly or releasing the gas phase of the drying medium too quickly,
4. a true critical point drying has not been performed, since the critical values for pressure and temperature were not achieved [according to Boyde and Maconnachie (1984) this is not a serious problem],
5. recondensation of the drying medium due to releasing the gas pressure in the drying chamber too quickly.

In addition, shrinkage effects may occur. They are, as a rule, dependent upon the type of chemicals used and have a uniform effect on the surface of the object. The reader is referred to Boyde and Maconnachie (1981) for a detailed discussion of these effects. More recent work by Boyde and Tamarin 1984, however, suggests that such shrinkages can be reduced considerably by using Freon 113 ($C_2Cl_3F_3$) as the intermediate medium. Similarly conventionally fixed specimens which have been treated with uranyl acetate solutions also tend to show reduced shrinkage (Wollweber et al. 1981).

Literature

Boyde A, Maconnachie E (1981) Morphological correlations with dimensional change during SEM specimen preparation. Scanning Electron Microsc 1981/IV: 27-34

Boyde A, Maconnachie E (1984) Not quite critical point drying. In: Revel J-P, Barnard T, Haggis GH (eds) The science of biological specimen preparation for microscopy and microanalysis. Scanning Electron Microsc Inc, AMF O Hare, Il, pp 71-75

Boyde A, Tamarin A (1984) Improvement to critical point drying technique for SEM. Scanning 6: 30-35

Fromme HG, Pfautsch E (1975) Die Trocknung wasserhaltiger Objekte mit der Kritischen-Punkt-Methode. In: Schimmel G, Vogell W (eds) Methoden-Sammlung der Elektronenmikroskopie. 7. Lieferung (Nr. 2.2.2) Wiss Verlagsges, Stuttgart, pp 1-19

Newbury DE, Joy DC, Echlin P, Fiori CE, Goldstein JI (1986) Advanced scanning electron microscopy and X-ray microanalysis. Plenum, New York, pp 1-454

Wollweber L, Stracke R, Gothe U (1981) The use of a simple method to avoid cell shrinkage during SEM preparation. J Microsc 121: 185-189

3.1.5.2 Freeze Drying

Both unfixed as well as chemically fixed objects can be freeze dried after freeze fixing (see Sect. 3.1.3.2). In order to do this, the frozen objects are placed on a cooled object table/holder in a vacuum chamber at a pressure of approximately 10^{-4} torr. The rate of ice sublimation is dependent upon the specimen temperature in the vacuum chamber (see also Chap. 2.6.4). At temperatures below $-130°$ C it is extremely low, thus for practical purposes one usually begins at a temperature of $-100°$ C and raises the temperature slowly to $-70°$ or $-60°$ C.

Under these conditions a cell monolayer can be freeze dried within 72 h, unfixed rat cartilage within 10 days (6 days at $-150°$ C, 2 days at $-100°$ C, 2 days at $-70°$ C). Before removing the objects from the vacuum chamber they should be brought to room temperature.

Many authors claim that freeze-dried objects show less shrinkage artefacts than those prepared by critical point drying. Often, however, one cannot discern any differences. Indeed freeze drying may produce superior results in the case of objects that contain a lot of water (e.g. vacuolate plant tissue, cartilage) provided that ice recristallisation can be avoided.

3.1.6 Mounting Specimens

Because of their hydrophilic character, dried specimens are best mounted and coated as soon as the drying procedure has been completed. Therefore one should prepare the specimen holders before the specimens become available. There are numerous different types of SEM specimen holder [a detailed description is given by Murphy (1982)], and there are many ways of sticking or "glueing" the specimens down onto them. However, a number of basic requirements have to be met in every case:

a) the "glue" should not be hygroscopic or wet the specimen, nor should it creep up (by capillary action) around the specimen and occlude portions of the specimen.
b) The "glue" should not lose its adhesive properties in the SEM; glued specimens should be able to withstand tilting.
c) The "glue" should make a good electrical contact with both specimen and holder.
d) The holder (and the "glue") should not emit backscattered electrons. Suitable materials for holders are therefore carbon and aluminium.

As a rule large objects, e.g. organs of animals, pieces of plants, or bones, can be stuck directly onto an aluminium holder with the help of a conducting "glue" (e.g. conducting silver or copper). For smaller objects the holder is first covered with a piece of double-sided adhesive tape and the object stuck down onto this. With the help of a binocular microscope the specimen can be oriented in such a way that the portion in which one is interested will be optimally exposed to the electron beam.

Small objects are best manipulated with eyelashes mounted on wood, glass or metal rods, somewhat larger objects with a hair brush. The objects are tipped out of the drying vessel onto a piece of filter paper and from there carefully manoeuvered onto the aluminium holder (covered with adhesive tape), which should be situated at the same level (see Fig. 3.4). In order to increase the conductivity one deposits a "bridge" of conducting silver/copper at the boundary between the adhesive tape and the surface of the holder. Microscopic objects such as pollen, microorganisms etc., can be added directly to the aluminium

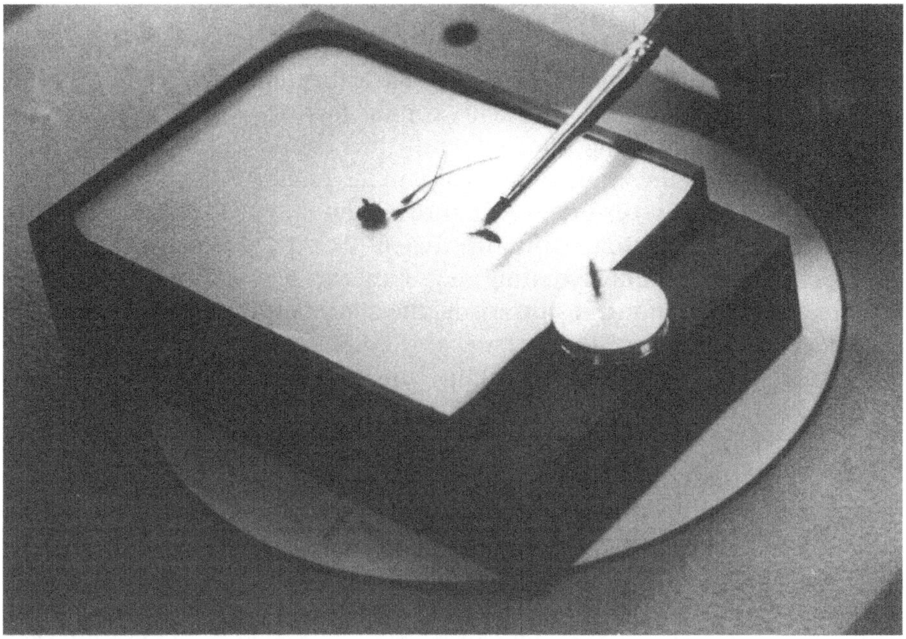

Fig. 3.4. Mounting dry specimens (in this case parts of an insect) onto an aluminium specimen holder

holder. Adhesion is achieved by scratching the surface of the holder beforehand, or by coating the surface with gelatine, formvar or polylysine.

Objects can also be mounted onto glass microscope slides which are then stuck down onto the aluminium holder (glass gives rise to a homogeneous background in the SEM). Membrane filters can also be mounted in this way. Objects grown on agar can be prepared by cutting out small blocks, fixing, dehydrating and drying them, and then mounting directly onto the holder.

In all of these cases it is important that the objects are not touched by fingers. This applies also to the attachment of the adhesive tape to the holder.

Literature

Murphy JA (1982) Considerations, materials, and procedures for specimen mounting prior to scanning electron microscopic examination. Scanning Electron Microsc 1982/II: 657-696

Witcomb MJ (1981) The suitability of various adhesives as mounting media for scanning electron microscopy. I. Epoxides, sprays and tapes. J Microsc 121: 289-308

Witcomb MJ (1985) The suitability of various adhesives as mounting media for scanning electron microscopy. II. General purpose glues. J Microsc 139: 75-114

3.1.7 Increasing Conductivity

In general, biological objects are poor electrical conductors or do not conduct at all (see Chap. 1.1.2.2). Conductivity can be imparted during the fixation (see Sect. 3.1.3.1) or by coating the surface of the object with a metal conductor. Both of these measures cause an increase in the number of secondary electrons emitted and lead to a considerable improvement in the video signal. In this respect a 1 nm-thick coating is often sufficient (for specimens with very uneven surfaces a 30–40-nm-thick coating may be necessary) and will often prevent charge accumulation at the surface of the specimen (recognizable as regular bright streaks and loss of sharpness – see Figs. 1.7 and 3.5). Moreover, the heating effects of the electron beam will be considerably reduced as a result of the coating.

Two coating methods exist: sputtering and vacuum evaporation (sublimation). In both cases heavy metals are involved, especially gold (Au) and gold-palladium (Au-Pd). Sometimes, however, carbon is deposited before the heavy metal. A major problem is the heating of the specimen which can result during the coating. Because of this one coats in several small steps, allowing the specimen to cool down between each step. Modern sputtering units reduce the thermal load during coating through various measures such as triode sputtering or "cold" diode sputtering which is done at 32° – 37° C. Sputtering is quicker to perform than a conventional vacuum sublimation and produces a more even

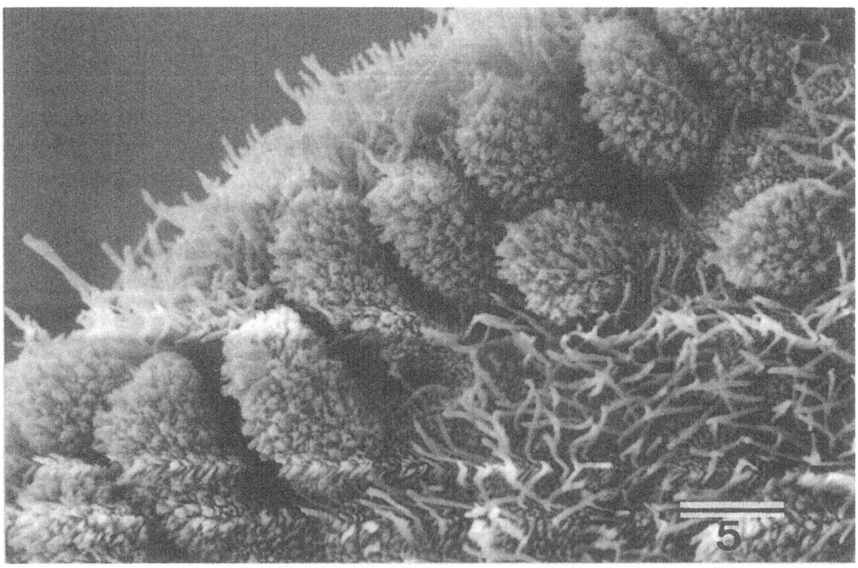

Fig. 3.5. A typical example of surface charge accumulation (object: anchoring organ of the plathyhelminth *Coelogynopora axi.* Fixed with glutaraldehyde and OsO_4, dehydrated through an alcohol series, critical point-dried with CO_2 but *not* sufficiently sputtered with gold). Magnification bar in μm

coating. In addition, because of their higher energy, heavy metal particles produced by sputtering adhere better to surfaces than do conventionally deposited metal layers.

Echlin (1978, 1981) has written detailed accounts of both coating methods and has concluded that a "cold" sputtering is superior for routine investigations. For high resolution coating in SEM see Newbury et al. (1986).

One final, but important point: the thickness of the coating should not exceed the resolving power of the SEM. In general, the thinner the coating the better is the detail which can be obtained.

Literature

Echlin P (1978) Coating techniques for scanning electron microscopy and X-ray microanalysis. Scanning Electron Microsc 1978/I: 109-132

Echlin P (1981) Recent advances in specimen coating techniques. Scanning Electron Microsc 1981/79-90

Newbury DE, Joy DC, Echlin P, Fiori CE, Goldstein JI (1986) Advanced scanning electron microscopy and X-ray microanalysis. Plenum, New York London, pp 1-454

3.1.7.1 ***Sputtering***

In this method the specimen and holder are brought into a cylindrical vacuum recipient. They are placed on a cooled base plate which functions as an anode. The cathode, which is gold-plated, is positioned above the specimen. After evacuating, the recipient is flooded with argon gas and again pumped out until a vacuum of $2\text{-}5 \times 10^{-4}$ bar is attained. A negative high voltage is then applied to the cathode; this results in a glow discharge which takes the form of a blue-violet plasma zone between the two electrodes. Both electrons as well as positively charged ions arise in this plasma zone. The latter accelerate and impinge upon the cathode, thereby releasing Au-atoms. These are forced in the direction of the anode and land either on the anode or the specimen. Should the Au-atoms collide with nonionized gas molecules they become deflected and hit the specimen at different angles. Thus parts of the specimen which are not directly facing the cathode will become coated; a uniformly thick Au-coating is the result.

Unfortunately with this method of coating, heating does become a problem, particularly due to electrons impinging upon the surface of the specimen. With the simple diode type of sputtering unit there is a cooling water supply around the anode and specimen. In triode types there is a ring-shaped electrode positioned above the cathode which attracts the majority of electrons. The "cold" diode type has a ring-shaped cathode which is placed around a centrally located permanent magnet. With this latter type of sputtering unit heat development is negligible.

Sputtering is very simple to perform when the makers' instructions are followed correctly.

Reproducibility can more or less be guaranteed providing the recipient contains a pure argon atmosphere. This necessitates exchanging the gas in the recipient several times before placing it under vacuum. If the thickness of the coating is not sufficient, one can always deposit more gold by a second sputtering.

3.1.7.2 Evaporating

With improved sputtering units coming onto the market, this method of coating has lost in importance for SEM. Nevertheless, it is to be preferred for certain purposes. Thus in element analysis it is often necessary to apply a carbon film to the specimen. This is usually achieved by thermal resistance heating, together with a rotating specimen stage, i.e. rotary evaporation (see also Chaps. 1.4.6, 2.3.1.3 and 2.6.5). Such specimens can be observed either directly in the SEM or after an additional heavy metal coating (made with either a sputtering unit or by evaporation). Heating problems are usually smaller when evaporating; however, the deposited coat is uneven and the chances of surface charge accumulation occurring are therefore higher.

Literature

Goldstein JJ, Newbury DE, Echlin P, Joy DC, Fiori C, Lifshin E (1981) Scanning electron microscopy and X-ray microanalysis. Plenum, New York London, pp 1-673

3.2 Storage of Specimens

There are a number of storage holders for SEM specimens available commercially.

Specimens are best stored in an exsiccator over silica gel or P_2O_5; here they can remain for years. Opening the exsiccator should, however, be done very carefully. If specimens are to be investigated within a few days of their preparation, they can be placed in dust-free containers in a dry room. In all events the specimens should not be touched with fingers, nor should they be breathed upon, both of which can lead to fractures in the metal coating in the vacuum of the SEM.

3.3 Demonstration of Surfaces via Replicas and Casts

Sometimes the specimen is too valuable, e.g. a museum object, too large (the removal of small parts damages the specimen), or so fragile (it will not survive the various preparatory steps) that it cannot be brought into the SEM. In such cases (and also when changes in surface structure during the development of one and the same specimen are to be investigated) a replica should be made of the surface in question. The replica, however, should fulfill the following criteria:

- it must not damage the specimen nor cause changes in its surface structures;
- it must not be toxic if the specimen is living and should remain living afterwards;
- it must faithfully reproduce the surface structures of the specimen;
- it should be easy to make and easy to release from the specimen;
- it must remain stable under the electron beam.

There are many methods for preparing replicas and there are numerous materials which can be used for them. Which type is most suitable can only be determined by scanning the literature cited at the end of this section.

Teeth and skin belong to the easiest and best types of object for the preparation of replicas. Silicone-rubber is usually used for the first (negative) replica and the second replica is most often from paraffin or acryl resins (very good results are obtained with a mixture of reprosil and Spurr's medium). If the replicas are nonconducting they must be mounted onto a specimen holder and coated.

For visualization of internal surfaces, e.g. of organs such as lung, kidney etc., a casting method is usually employed. Mercox, methyl-methacrylate or Batson No. 17 (see Lametschwandtner et al. 1984) is injected into the organ to be investigated and then hardened. The vascular system of these organs can then be revealed after removing the biological material with concentrated acids or alkali.

Literature

Gannon LBJ (1978) Vascular casting. In: Hayat MA (ed) Principles and techniques of scanning electron microscopy, vol VI. Biological applications. Van Nostrand-Reinhold, New York, pp 170-193

Lametschwandtner A, Lametschwandtner U, Weiger T (1984) Scanning electron microscopy of vascular corrosion casts - technique and applications. Scanning Electron Microsc 1984/II: 663-695

Murakami T (1978) Methyl methacrylate injection replica method. In: Hayat MA (ed) Principles and techniques of scanning electron microscopy, vol VI. Biological applications. Van Nostrand-Reinhold, New York, pp 159-169

Pameijer CH (1975) Replica techniques. In: Hayat MA (ed) Principles and techniques of scanning electron microscopy, vol IV. Biological applications. Van Nostrand-Reinhold, New York, pp 45-93

Pameijer CH (1979) Replication techniques with new dental impression materials in combination with different negative impression materials. Scanning Electron Microsc 1979/II: 571-574

Rosenbauer KA, Kegel BH (1978) Rasterelektronenmikroskopische Technik. Präparationsverfahren in Medizin und Biologie. Thieme, Stuttgart, pp 1-241

3.4 Visualization of Internal Surfaces Through Sectioning and Dry-Fracturing (Dry-Cleaving)

SEM of biological objects is not restricted to external surfaces; it is also possible to investigate the surfaces of internal structures. Two methods are available for this; one involves the preparation of sections from hard or hardened material, the other entails fracturing frozen or dried material (dry-fracturing or dry-cleaving). In the former, the surfaces which are to be exposed can be chosen; in the latter case, the object breaks open randomly at one or many levels, resulting in a very high degree of plasticity (e.g. Fig. 3.1 b).

If an object is hard, e.g. wood, there is no difficulty in preparing sectioned surfaces. With animal tissue this is best done after fixing and dehydration. If necessary, the specimens can be embedded in plastic (see Chap. 2.1.3), and after complete partial removal of the embedding material, the exposed surfaces of sectioned material can be viewed in the SEM.

Dry-fracturing is comparatively easy to carry out on dry material. The object is mounted onto double-sided adhesive tape which is stuck onto a specimen holder. A second piece of adhesive tape is carefully pressed down onto the object. This piece of tape is then ripped off, taking some of the object with it. It, also, can be stuck down onto a specimen holder, allowing complimentary fractured surfaces to be seen. The procedure can be repeated several times, allowing a number of internal surfaces in one and the same object to be visualized.

Inner surfaces can be produced from moist material by freeze fracturing. According to Humphreys et al. (1974, 1978), one should proceed as follows:

A cylinder of parafilm is made by winding a 2-cm-wide strip of parafilm around a rod of 2 mm diameter. Fixed, and dehydrated (in 100% ethanol) objects are inserted into the cylinder. The cylinder is sealed at both ends, held with fine forceps and dipped into liquid nitrogen. The frozen cylinder is then placed onto a metal block (precooled to liquid N_2 temperature) and fractured open with a precooled single-edged razor blade. The fractured pieces are then transferred to absolute ethanol. After removing the parafilm fragments, the fractured objects are dried with the critical point method.

Literature

Flood PR (1975) Dry-fracturing techniques for the study of soft internal biological tissues in the scanning electron microscope. Scanning Electron Microsc 1975/I: 287-294

Goldstein JJ, Newbury DE, Echlin P, Joy DC, Fiori C, Lifshin E (1981) Scanning electron microscopy and X-ray microanalysis. Plenum, New York London, pp 1-673

Humphreys WJ, Spurlock BO, Johnson JS (1978) Critical-point drying of cryofractured specimens. In: Hayat MA (ed) Principles and techniques of scanning electron microscopy, vol VI. Biological applications. Van Nostrand-Reinhold, New York, pp 136-158

Laane MM (1976) Sectioned specimens. In: Hayat MA (ed) Principles and techniques of scanning electron microscopy, vol V. Biological applications. Van Nostrand-Reinhold, New York, pp 36-52

Reimer L, Pfefferkorn G (1977) Raster-Elektronenmikroskopie, 2nd edn. Springer, Berlin Heidelberg New York, pp 1-282

3.5 Element Analysis

In the normal mode of operation SEM can reveal many structural details, but it is difficult to make inferences on the functional aspects of the objects under investigation. A step in this direction can, however, be made by taking advantage of the phenomenon of cathodoluminescence and by using X-ray microanalysis. Both of these methods enable one to learn something of the nature of the material one is studying and how the elements in it are distributed. One should note that these methods are neither restricted to biological objects nor are they purely scanning methods - they can also be used in TEM [see Reimer and Pfefferkorn (1977) for theory and further details].

In cathodoluminescence, electrons which impinge upon an object cause a luminescence. This is, in principle, the same kind of excitation response that occurs in optical fluorescense microscopy. The degree to which the object "luminesces" is dependent upon its ability to do so and on the concentration of the elements in the object responsible for this effect. The amount of light emitted is, in the case of biological objects, very small. As a result, special detectors (see Herbst and Hoder 1978) capable of collecting the light over a wide angle are necessary. The light is then transmitted via a light cable to a photomultiplier. Since the amount of light emitted is inversely proportional to temperature, it is usual to investigate specimens at low temperatures.

The native or inherent fluorescence capability of biological material can be successfully exploited to visualize chalk deposits in tissues or sclerified walls in blood vessels. Treatment with conventional fluorochromes leads to an additional excitation (see Fig. 3.6), permitting a differentiation of the tissue from a physiological point of view. Fluorescent-tagged antibodies can also be applied in this way. A problem which can arise, however, relates to the relatively short life-time of some organic fluorescent chemicals. For example, acridine orange and quinacrine lose their fluorescent properties very quickly upon exposure to the electron beam. Others, such as coriphosphine or sodium fluorescein, are much more stable.

The second method, dispersive X-ray microanalysis, puts to use the fact that atoms, when hit by electrons, emit X-rays which are characteristic for the particular atom involved. Electrons falling to lower energy levels lead to an intermittent type of energy release of precisely defined qualities. Line spectra arise corresponding to the particular atoms involved. Thus information on the types of element, their concentration and their distribution can be obtained by analyzing such X-ray spectra.

The spectra can be produced with the help of either a wavelength dispersive spectrometer or an energy dispersive spectrometer. In the former, the X-rays are diffracted by a crystal grating which spreads the rays out according to their

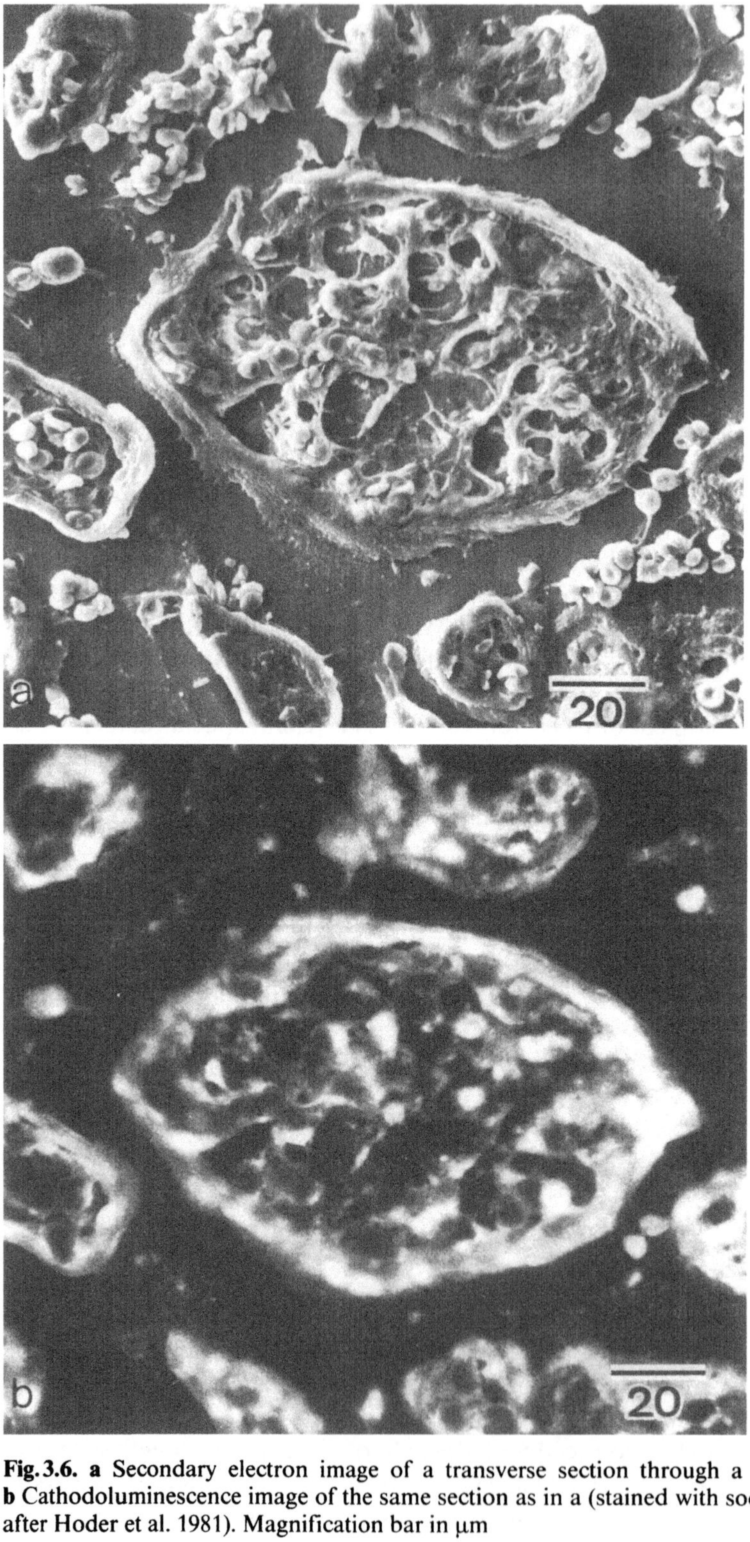

Fig. 3.6. a Secondary electron image of a transverse section through a human placenta. **b** Cathodoluminescence image of the same section as in a (stained with sodium fluorescein; after Hoder et al. 1981). Magnification bar in μm

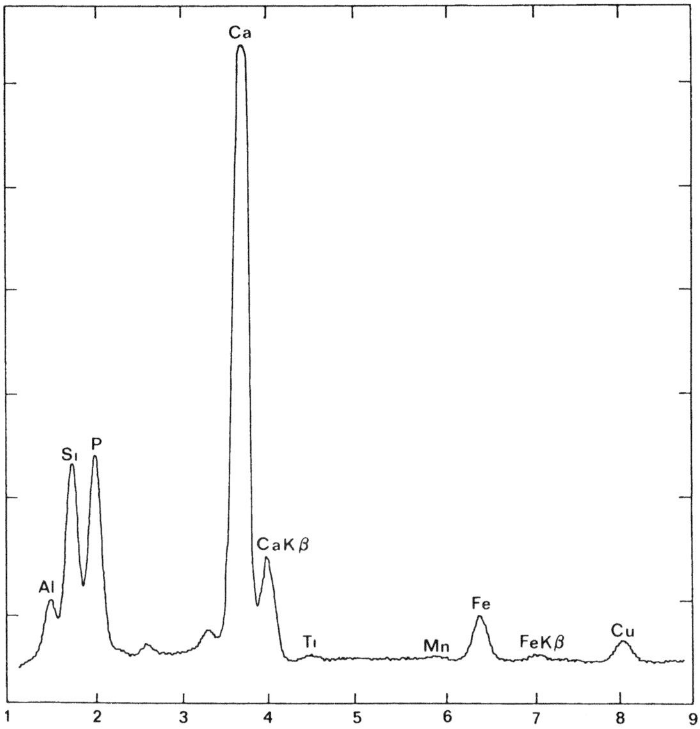

Fig. 3.7. Total emission spectrum of a specimen. Prehistoric human bone (buried) with incorporations of the matrix. *Ordinate* number of impulses registered. *Abscissa* energy of the emitted radiation (in keV; K_α radiation except where stated)

wavelength. More frequently in use is the energy dispersive system (e.g. ORTEC, EDAX) consisting of a cooled semi-conductor detector, based on Si and Li, which is fitted onto the specimen chamber of the microscope. The energy of the emitted X-rays is determined simultaneously for all elements and the values printed out graphically (e.g. Fig. 3.7). With such systems one can detect and discern those elements lying between Li(3) and U(92). As a rule the spectra allow one to make qualitative, rather than quantitative, statements about elements present in an object, although this is also possible. In this respect one should note that individual specimens may vary by as much as 10% from one another.

Two restrictive features of dispersive X-ray microanalysis are the facts that

a) the high accelerating voltages involved may lead to considerable heating of the specimen;
b) the possibility of detecting elements is not the same for all elements even if they are present in the same amounts in a specimen. Thus the specimen should be investigated at various accelerating voltages: for Al the excitation optimum lies around 5 kV, for Fe it is between 10 and 15 kV, and 20–30 kV is suitable for the majority of the heavy metals.

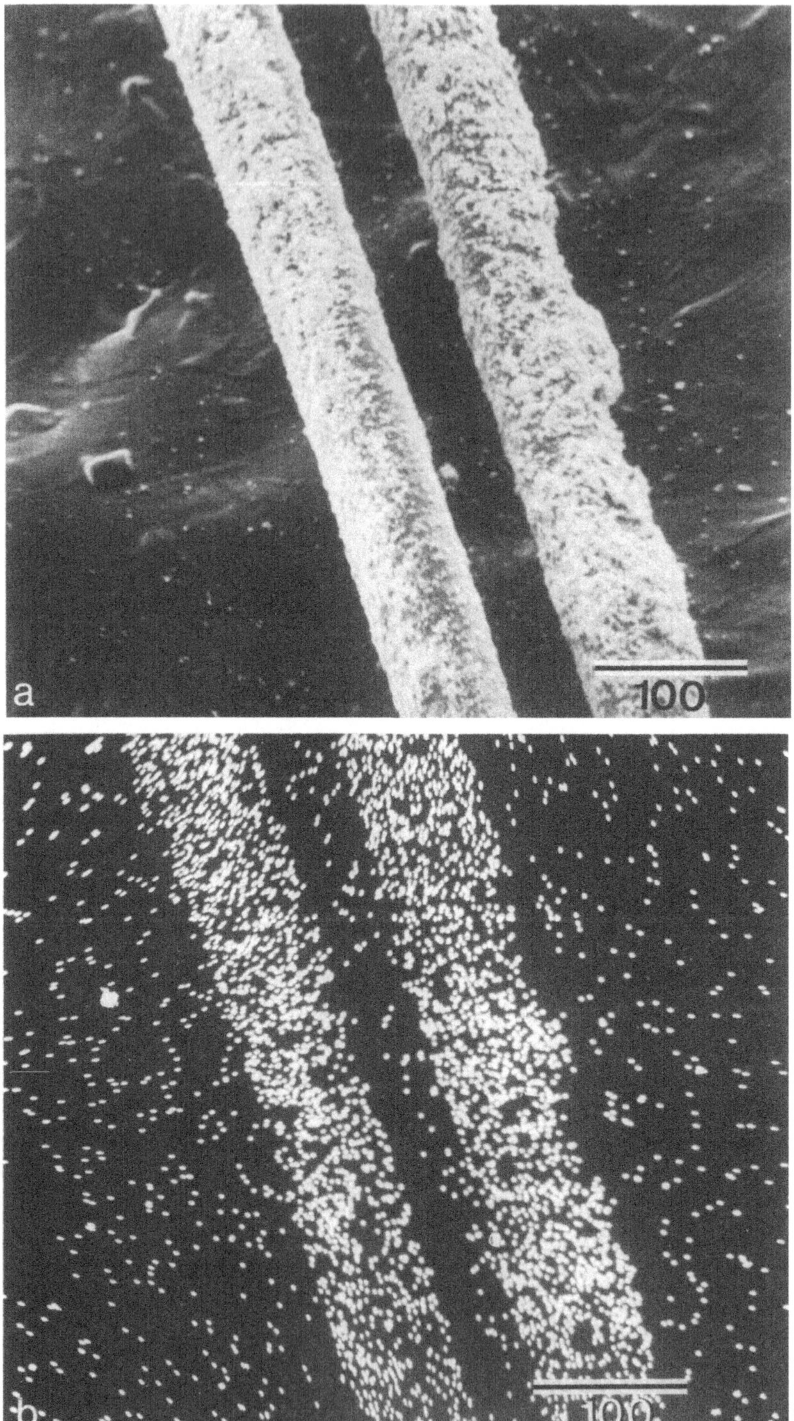

Fig. 3.8. a Secondary electron image of two cinnabar-(HgS)-coloured head hairs of a South American mummy. **b** Element distribution image of mercury (Hg). (Micrographs Hermann and Staubach). Magnification bar in μm

In addition to the possibility of performing an element analysis without destroying the specimen, the energy dispersive system has an additional advantage. The X-ray emission which occurs as a result of the scanning electron beam can be converted on the viewing screen into a series of points. The distribution of elements in an object can therefore be visualized optically as well as graphically. This is also possible for individual elements (e.g. Fig. 3.8), enabling once again a functional correlation to be made with the structures viewed.

With respect to the preparation of a specimen for element analysis the reader is referred to the standard work of Goldstein et al. (1981). In general, one must be aware of the various qualitative and quantitative changes in the amount and distribution of elements which can possibly occur during the preparation. This is especially so during the fixation, washing and dehydration steps. Also important is that elements should not become masked by chemicals used in the preparation. Embedding may also be a problem, particularly when changes in form result. Finally, conductivity is normally obtained with a carbon coat that does not interfere with the X-ray emission and leaves a clear background.

Literature

Goldstein JI, Newbury DE, Echlin P, Joy DC, Lifshin E (1981) Scanning electron microscopy and X-ray microanalysis. Plenum, New York London, pp 205–495

Herbst R, Hoder D (1978) Cathodoluminescence in biological studies. Scanning 1: 35–41

Hoder LD, Herbst R, Multier-Lajous A-M (1980) Farbverschiebungen in Kathodolumineszens-Aufnahmen. BEDO 13: 153–156

Newbury DE, Joy DC, Echlin P, Fiori CE, Goldstein JI (1986) Advanced scanning electron microscopy and X-ray microanalysis. Plenum, New York London, pp 1–454

Reimer L, Pfefferkorn G (1977) Raster-Elektronenmikroskopie, 2nd edn. Springer, Berlin Heidelberg New York, pp 186–222

4 Evaluation of Micrographs

4.1 Morphometry

4.1.1 Problems and Solutions

Often a statement about a particular object, made as a result of an electron microscope examination, can only become convincing when backed up with numerical data obtained by image analysis. In addition, a statistical treatment of the measurements made can sometimes give rise to information which is not readily apparent in a qualitative description of the object concerned. Thus, morphometry entails methods of measurement as applied to morphological problems. An important area of morphometry is stereology, which provides the mathematical framework to enable one to create three-dimensional models from measurements carried out on two-dimensional images. Micrographs of sections, freeze-fracture replicas or freeze- or critical point-dried preparations often contain complicated spatial structures to which stereological methods must be applied.

Many particles such as molecules, viruses and bacteria can be characterized on the basis of linear measurements. However, in addition to determinations of length, area, or volume, frequently occurring tasks in morphology are the comparison of corresponding structures in related objects, the description of the development of a particular structure and the analysis of changes in a structure in relation to biochemical or physiological experimentation.

4.1.2 Measurement: Some General Points

Forms such as spheres, ellipsoids, cylinders or cubes are often encountered in bacteria, viruses and molecules and allow area, volume and weight to be calculated from linear measurements. For objects which have a more complicated shape (this applies equally to both small and large structures), the most reliable information is that obtained with the help of serial sectioning, although with the help of stereology it is possible to deduce length, area and volume from objects seen in single sections.

Here we will attempt to give a summary of the most important methods; for more details the reader is referred to the literature at the end of this section.

An important prerequisite in morphometric determinations is the calibration of magnification steps in EM negatives (and, of course, positives). Sometimes true magnifications may vary up to 10% from those values read off the microscope. Again, an internal standard of known size, e.g. latex particles, tobacco mosaic virus or ferritin, which can be included in the micrograph, is of great value.

Another prerequisite in morphometric determinations is a knowledge of all potential sources of artifact as well as of the limits of the particular method of measurement. If an internal standard cannot be included, the extent to which the method of preparation can change the dimensions of the object must be known (e.g. what sort of an effect does spreading have on the length of nucleic acid molecules?). In the morphometry of cells and tissues a factor must be introduced to compensate for potential shrinkages which have occurred during fixing and dehydration. Similarly, the extent or thickness of the coating must be taken into account when dealing with replicas, shadow-cast or SEM specimens. In general, methods of preparation should be standardized as much as possible in order to reduce the number of factors which have to be corrected for.

4.1.3 Stereology: General Principles

Stereological methods must be employed when a serial section analysis cannot be carried out and the volume, area etc., of an object is to be determined. Numerous books are available on this complex subject (see literature list) in which detailed descriptions of the various methods are given.

Before a stereological analysis can be carried out, the micrographs must be checked regarding a number of aspects. Firstly the structures to be investigated must be present in sufficiently large numbers, of uniform size, and they must be easy to recognize in section. Secondly it is assumed that the number of sectional profiles of a structure visible in a particular section is directly related to the total amount of these structures in the object being investigated. It follows then that the sectional profiles must be statistically present in the same number irrespective of the orientation of the section. Whereas this might be true for structures with a homogeneous composition, it is not necessarily so for anisotropic structures. Unfortunately, such structures are often encountered in biological objects, e.g. nerve cells and muscle fibres, and require the application of special stereological methods which have been developed for such purposes.

4.1.4 Collection and Evaluation of Data; Statistical Treatments

An $(n-1)$-dimensional image is presented in a sectional profile of an n-dimensional structure. Hence a spherical object is seen as a flat profile (in this case a circle), surfaces are represented as lines, and rodlike objects as pointlike structures. In general, the simpler the method used to measure and record these contours the more reliable is the final 3-D reconstruction. This is simply because a larger number of measurements can be made in a given period leading to a greater degree of statistical reliability.

Measurements can be performed manually and the data used to calculate the form/shape/volume of the structure involved; alternatively this can be done with the help of electronic instruments coupled to a computer. The manually performed collection of data usually involves placing a transparent plastic foil, with an appropriate pattern on it, onto the micrograph to be evaluated. The pattern varies depending on the type of measurement to be made, but can be a linear, or point, or grid system, or can be a combination of various regularly placed symbols or lines. Foils can be constructed in the laboratory or can be obtained commercially.

The computer-assisted evaluation of morphometric data has superceded that by the manual method. Specially constructed apparatuses with a standard, built-in-programme (e.g. MOP-system, Kontron GmbH) are now available. Just as useful, indeed because of the flexibility involved in terms of programming in some ways superior, is a self-assembled system comprising a small computer and a graphics tablet (e.g. the Apple system).

In some cases, e.g. with the IBAS system (Zeiss GmbH), it is possible to carry out morphometric determinations directly on the screen in the microscope. It is also possible (with Zeiss electron microscopes and the Morphomat 10) to incorporate a graphics tablet and with the help of a light point (projected onto the observation screen via a mirror), one can describe the sectional profile of an object, which is then computed.

Statistical certainty increases with the number of individual measurements. The arithmetical average, x, of a particular number, n, of a group of measurements can be expressed according to the following formula:

$$\bar{x} = \frac{1}{n}(x_1 + x_2 + \ldots\ldots x_n) = \frac{\sum x}{n}.$$

The standard deviation s is expressed as follows:

$$s = \sqrt{\frac{(x - \bar{x})^2}{n-1}}.$$

For additional formulae in statistics the reader is referred to the various textbooks listed at the end of this chapter.

Literature

Agar AW, Alderson RH, Chescoe D (1974) Principles and practice of electron microscope operation. In: Glauert AM (ed) Practical methods in electron microscopy. Elsevier/North Holland, Amsterdam, pp 1-345

Bailey NTJ (1959) Statistical methods in biology. English Univ Press, London, pp 1-379

Ferguson J, Davis RW (1978) Quantitative electron microscopy of nucleic acids. In: Koehler JK (ed) Advanced techniques in biological electron microscopy, vol II. Springer, Berlin Heidelberg New York, pp 123-171

Gilbert N (1973) Biometrical interpretation. Oxford Univ Press, London, pp 1-125
Haug H (1976) Die verschiedenen Verfahren zur Werteerfassung in der biologischen Morphometrie und Stereologie. Microsc Acta 78: 197-220
Hoppe W (1977) Strukturanalyse mit Elektronenstrahlen (Elektronenmikroskopie). In: Hoppe W, Lohmann W, Markl H, Ziegler H (eds) Biophysik. Springer, Berlin Heidelberg New York, pp 67-87
Mishima Y, Matsunaka M (1976) Quantitative and three-dimensional analysis of dendritic cells and keratin cells in the human epidermis revealed by transmission and scanning electron microscopy. In: Yamada E, Mizurda V, Kurosima K, Nagano T (eds) Recent progress in electron microscopy of cells and tissues. Thieme, Stuttgart, pp 290-304
Weibel ER (1969) Stereological principles for morphometry in electron microscopic cytology. Int Rev Cytol 26: 235-302
Weibel ER (1979a) Stereological methods, vol I. Practical methods for biological morphometry. Academic Press, London New York, pp 1-415
Weibel ER (1979b) Stereological methods, vol II. Theoretical foundations. Academic Press, London New York, pp 1-340
Willison JHM, Rowe AJ (1980) Resolution and quantitation in replicas and shadowed specimens. In: Glauert AM (ed) Practical methods in electron microscopy, vol VIII. Elsevier/North Holland, Amsterdam, pp 245-282

4.2 Averaging and Image Reconstruction

4.2.1 General

There are inherent difficulties in making deductions about the true structure of biological macromolecules on the basis of EM images. One reason for this is the possibility that the form visualized in the EM includes artifactual changes occurring as a result of one or more of the preparatory steps, e.g. through dehydration or drying down onto the support film or through the damaging effect of the electron beam. Another factor is the reduced resolution due to staining with heavy metal salt solutions. Nevertheless, despite these potential drawbacks, one can extract much information which can be critically analyzed. A final goal in macromolecular electron microscopy, for example, is to extract the common features present in numerous images of one and the same molecule and to construct a model which incorporates all of these. There are a number of methods which enable one to do this; all of them constitute essentially an averaging of the extractable data.

4.2.2 Markham Rotation

Isolated protein molecules often give rise to images in the EM which suggest a certain degree of symmetry (see Fig. 4.1 a). Frequently this is a rotation symmetry and a relatively simple photographic averaging method can be employed to test this.

The first step is the production of an enlarged second negative in which the structures to be analyzed appear bright against a dark background. This negative is then inserted into the enlarger and so positioned that the structure to be investigated lies in the centre of the light beam. A magnification should be chosen which gives rise to an image (i.e. on the positive) of at least 1.5 cm diameter. Assuming that a sixfold rotational symmetry is to be expected, an exposure is then made for about ¼ to ⅙ of the optimal exposure time. The positive paper is then rotated on its axis (the axis parallel to the light beam) by ⅙th of 360°, i.e. 60°. This is repeated five times (six exposures in all) with the same piece of positive paper, until the start position is reached. If indeed a sixfold symmetry was present, the structure will appear in a "reinforced" form when the positive is developed, and should be much easier to recognize as such in comparison to the nonrotated positive with a normal exposure. In order to rule out other symmetry possibilities, control exposures with other angles (e.g. ⅓rd of 360°, ⅕th of 360°, ⅐th of 360°) should be made. If additional symmetry possibilities are suspected it is advisable to carry out exposures corresponding to 2-, 3-, 4-, 5-, 7-, 8- and 12-fold rotations. Finally, the result obtained should correlate well with the available biochemical data on the number of subunits.

Literature

Antranikian G, Klinner Ch, Kümmel A, Schwanitz D, Zimmermann Th, Mayer F, Gottschalk G (1982) Purification of L-glutamate-dependent citrate lyase from *Clostridium sphenoides* and electron microscopic analysis of citrate lyase isolated from *Rhodopseudomonas gelatinosa, Streptococcus diacetilactis* and *C. sphenoides*. Eur J Biochem 126: 35-42

4.2.3 Principles of Light Optical Diffraction

An optical diffraction can be made of an electron micrograph providing that there are periodic structures with sufficient contrast to be seen (e.g. Fig. 4.1 b). Necessary for this procedure is an optical bench equipped with a laser, camera and lenses suitable for accommodating diffracted light. The negative is inserted in the laser beam; if the region with the periodic structure is smaller than the beam diameter it must be bordered off with a circular or quadratic mask. The laser beam is diffracted by the periodic structures in the negative and the diffraction pattern, in the form of a series of regularly ordered spots (see Fig. 4.1 c), is recorded by the camera. From such a diffractogram one can obtain information as to spacings and angles in the original structure. Diffraction spots, which are diffuse, correlate with poorly ordered areas in the negative - these may be a result of deformed areas of the object or could simply represent the background (i.e. support film). However, such diffraction spots can also be used to determine the relative degree of focus of the negative (Fig. 4.1 d). If the ring formed by these diffraction spots is not circular, the negative is in addition astigmatic (Fig. 4.1 c).

If one wants to separate the diffuse background from the important part of the diffractogram, the image can be reconstructed and the background filtered

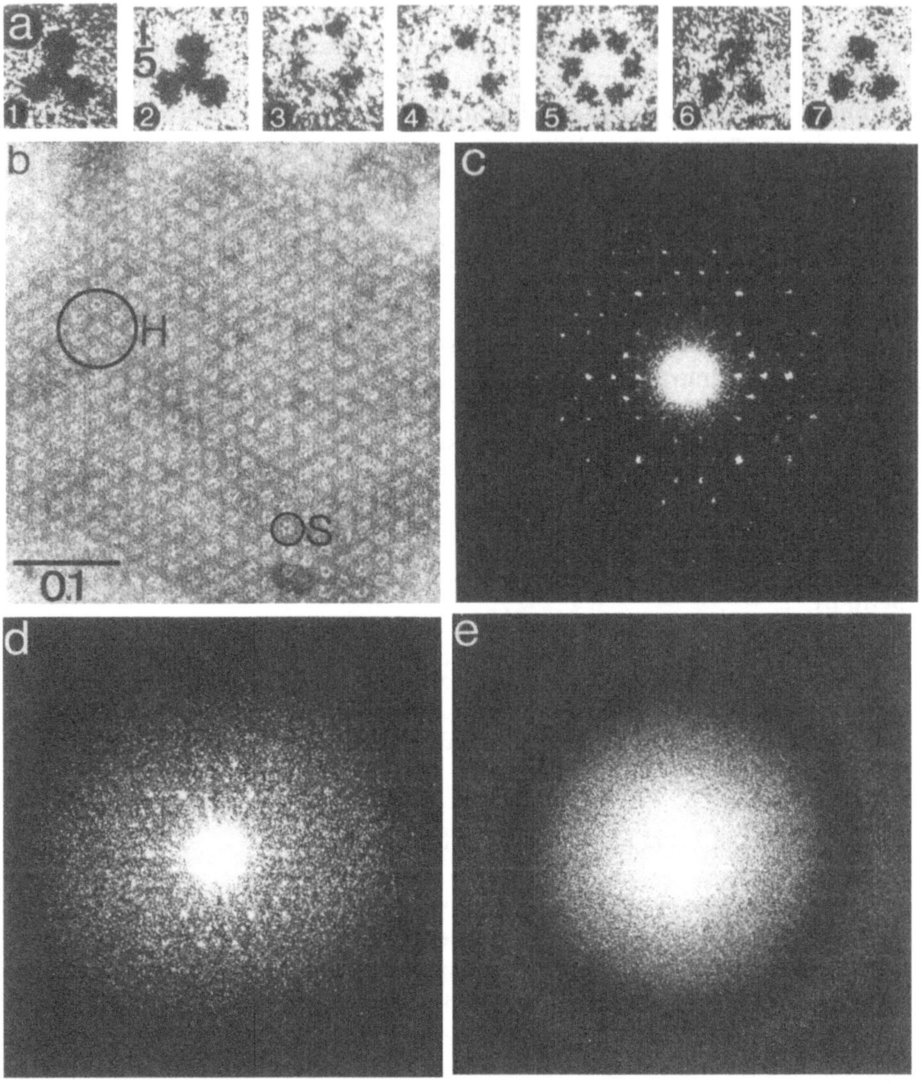

Fig. 4.1. a Electron micrographs of the various forms of the enzyme citrate lyase (see also Figs. 2.21 c and 2.26 k) *(1, 3, 6)* as well as the results of a Markham rotation image analysis *(2, 4, 5, 7)*. *1* "star"; *3* "ring"; *6* intermediate form. *2, 4, 7* rotation in three steps, each 120°; *5* rotation in six steps, each 60°. Averaging of mass densities as a result of this procedure is particularly clearly demonstrated. *5* is not permissible since in the original micrographs *(3)* an alternating circular pattern of large and small subunits is recognizable. If image *5* were true there would be no difference in the size of the subunits. Negative staining with uranyl acetate. Magnification bar in μm (Antranikian et al. 1982).
b Negatively stained (with uranyl acetate) isolated fragment of the "surface layer" of a hydrogen oxidizing bacterium. The structures in the layer are hexagonally ordered *(H)*; each structure *(S)* consists of six subunits, which are also hexagonally arranged. In addition there is a fine periodic structure observable in the background. Magnification bar in μm.
c–e Optical diffractograms of the image shown in **b**; **c** diffraction points of various orders and intensities are clearly visible and prove the presence of a hexagonal structure; **d** as for **c** but

out. To do this the optical bench is extended and an extra lens system is incorporated, comparable to that used to collect and image the diffracted rays, but in the reverse order. In place of the camera a mask is inserted with holes punched in it at the positions of the various diffraction maxima as well as one for the central, undiffracted, beam. The image which is finally produced is thus free from background interference, and is usually much clearer than the original EM negative. If the negative used is that of a negatively stained helicoidal, planar or other kind of complex structure, a one-sided image reconstruction can be performed by selecting the appropriate diffraction maxima.

An optical diffractometer is routinely used to determine which negatives are suitable for computer-assisted image reconstruction (see below).

Literature

Horne RW, Markham R (1973) Application of optical diffraction and image reconstruction techniques to electron micrographs. In: Glauert AM (ed) Practical methods in electron microscopy, vol I, part 2. Electron diffraction and optical techniques. Elsevier/North Holland, Amsterdam, pp 325–435

4.2.4 Principles of Computer-Assisted Image Reconstruction

This is not a technique which belongs to the usual methods arsenal of a cytology laboratory. Not only does it require considerable experience in high resolution electron microscopy, it demands a lot of mathematical expertise. A computer-assisted image reconstruction can be made in two or three dimensions, depending on the information available. The object can be an isolated macromolecule or a periodically structured object such as a protein crystal, a membrane with regularly arranged substructures, microtubules etc. For a two-dimensional reconstruction only negatives showing one projection of the object are necessary; for 3-D reconstructions the object has to be photographed systematically several times as a tilting series (see Chap. 2.3.1.5).

The information in the negative must be transformed into a form which can be given to a computer. This is usually done by scanning the negative with a microdensitometer which records numerically the relative "blackness". In the case of isolated macromolecules this is carried out for a number of individual imaged molecules in order to arrive at an average.

For 2-D reconstructions the first task is to look for the smallest repeating unit, the unit cell. Averaging is then done for such an area. With the appropriate

◁

with the diffractogram at a lower magnification. Because of a different exposure the diffraction points are now covered with unspecific noice; the diffractogram is now not exactly circular thus indicating that the original micrograph was astigmatic; **e**, as for **c** and **d**. The almost perfect circular form of the diffractogram indicates that in the original micrograph little or no astigmatism was present

experimentation, such images can be used to determine the mass density of the object at any particular position. Thus, depending on the method of preparation and the type of staining, one can resolve structural details down to 1 nm.

3-D reconstructions are much more difficult and require an appropriate computer programming so that a comparison of the information for the various image projections can be made. A 3-D density distribution profile is normally the result, which, if necessary, can be printed out sectionally.

Literature

Baumeister W, Vogell W (eds) (1980) Electron microscopy at molecular dimensions; state of the art and strategies for the future. Springer, Berlin Heidelberg New York, pp 1-353

Appendix: Buffers in Electron Microscopy

Buffers should be prepared keeping in mind the pKa value (see Chap. 2.1.1.3.1) as well as the temperature. The appropriate pH can be obtained either by titrating acid or base against the buffer solution or by mixing together volumes according to a table. The following buffers are the ones most usually used in biological electron microscopy. Please note that for concentrations other than 0.1 M different amounts must be weighed out.

1. Sörensen's phosphate buffer (0.1 M)

Solution A: 0.1 M KH_2PO_4 (13.76 g l^{-1})
Solution B: 0.1 M $Na_2HPO_4 \cdot 2H_2O$ (18 g l^{-1})

Desired pH at 25° C	Mix together to make 100 ml	
	A ml	B ml
6.0	87.7	12.3
6.8	50.8	49.2
7.0	39.2	60.8
7.2	28.5	71.5
7.4	19.6	80.4
7.6	13.2	86.8
7.8	8.6	91.4
8.0	5.5	94.5

2. Sodium phosphate buffer (0.1 M)

Solution A: 0.1 M $NaH_2PO_4 \cdot 2H_2O$ (15.6 g l^{-1})
Solution B: 0.1 M $Na_2HOP_4 \cdot 2H_2O$ (18 g l^{-1})

Desired pH at 25° C	Mix together to make 100 ml	
	A ml	B ml
6.0	87.7	12.3
6.8	51.0	49.0
7.0	39.0	61.0
7.2	28.0	72.0
7.4	19.0	81.0
7.6	13.0	87.0
7.8	8.5	91.5
8.0	5.3	94.7

3. Cacodylate buffer (0.1 M)

Solution A: 0.2 M $Na_2(CH_3)_2AsO_2 \cdot 3H_2O$ (42.8 g l^{-1})
Solution B: 0.2 M HCl 16.6 ml conc. (36–38%) HCl in 1 l H_2O

Desired pH at 25° C	Mix together to make 100 ml		
	A ml	B ml	H_2O ml
6.4	50	36.8	13.2
6.6	50	26.8	1.2
6.8	50	18.8	32.1
7.0	50	12.8	37.2
7.2	50	8.4	41.6
7.4	50	5.6	44.4

4. Sulfonic acid ("Good") buffers

Buffer	pKa	Buffering range	Set pH with
HEPES (N-2-hydroxy ethylpiperazine-N′-2-ethanesulfonic acid)	7.5	6.8–8.2	NaOH/HCl
MOPS (3-N-morpholino-propane sulfonic acid)	7.2	6.5–7.9	NaOH/HCl
PIPES (Piperazine-N-N′-bis ethane sulfonic acid)	6.8	6.1–7.5	NaOH/HCl

Subject Index

Numbers in *italics* refer to figures

GPSR Compliance

The European Union's (EU) General Product Safety Regulation (GPSR) is a set of rules that requires consumer products to be safe and our obligations to ensure this.

If you have any concerns about our products, you can contact us on ProductSafety@springernature.com

In case Publisher is established outside the EU, the EU authorized representative is:

Springer Nature Customer Service Center GmbH
Europaplatz 3
69115 Heidelberg, Germany

Batch number: 10607330

Printed by Printforce, the Netherlands